GREENWOOD GUIDES TO
BIOMES OF THE WORLD

Introduction to Biomes
Susan L. Woodward

Tropical Forest Biomes
Barbara A. Holzman

Temperate Forest Biomes
Bernd H. Kuennecke

Grassland Biomes
Susan L. Woodward

Desert Biomes
Joyce A. Quinn

Arctic and Alpine Biomes
Joyce A. Quinn

Freshwater Aquatic Biomes
Richard A. Roth

Marine Biomes
Susan L. Woodward

Temperate Forest Biomes

Temperate Forest
BIOMES

Bernd H. Kuennecke

Greenwood Guides to Biomes of the World
Susan L. Woodward, General Editor

GREENWOOD PRESS
Westport, Connecticut • London

Library of Congress Cataloging-in-Publication Data

Kuennecke, Bernd.
    Temperate forest biomes / Bernd H. Kuennecke.
        p. cm. — (Greenwood guides to biomes of the world)
    Includes bibliographical references and index.
    ISBN 978-0-313-33840-3 (set : alk. paper) — ISBN 978-0-313-34018-5 (vol. : alk. paper)
    1. Forest ecology—North America. 2. Temperate climate. 3. Mediterranean climate. 4. Biotic communities. I. Title. II. Series.
    QH102.K84 2008
    577.3097—dc22   2008024625

British Library Cataloguing in Publication Data is available.

Library of Congress Catalog Card Number: 2008024625
ISBN: 978-0-313-34018-5 (vol.)
        978-0-313-33840-3 (set)

First published in 2008

Greenwood Press, 88 Post Road West, Westport, CT 06881
An imprint of Greenwood Publishing Group, Inc.
www.greenwood.com

Printed in the United States of America

♾™

The paper used in this book complies with the Permanent Paper Standard issued by the National Information Standards Organization (Z39.48–1984).

10 9 8 7 6 5 4 3 2 1

Contents

Preface

As the author, an admittedly subjective reason for undertaking this particular project has been my significant personal stake in this volume. My personal and professional interests in forests date back more than 40 years. Travel through various portions of the forested areas of Europe (Germany, Belgium, Luxembourg, France, Italy, Austria, Switzerland, Denmark, Norway, and Sweden) was undertaken not only in conjunction with vacations, but also in response to my curiosity of land usages of forests regions. When working on a dissertation on the effects of primary industries in Oregon, I investigated the vegetation associations of that state, as well as the effects that had been caused by the human use of such natural resources. These interests were continued and deepened, in conjunction with a series of summer field trips into the forest biomes described in this volume, for both pleasure as well as in preparation for subsequent university-sponsored field trips and teaching of field research methods courses. Extensive travel was undertaken in North America with university students while teaching field research methods classes. These research trips provided numerous opportunities to repeatedly visit the forest regions of the Pacific Northwest, western Canada, Alaska, northeastern Canada, the New England region, and Appalachia. Annual trips to Europe and repeated visits to the forested regions of Northern Germany (with particular interests concerning the differences in land usages of forested areas in the former East and West Germany) have been ongoing for some 30 years to date.

The biomes described in this volume are accessible for direct inspection by the reader in numerous areas of the world. Travel into and through the geographic expressions of the biomes that are described in the following chapters will provide the reader with such direct impression. For those living within travel distance, some easy access routes are described. Boreal forest regions are accessible to those who can travel in southeastern Canada, Alaska, northern Michigan, the northern parts of the New England states, the mountainous areas of western Canada and the Pacific Northwest of the United States, as well as along the high ridges of the northern and central Appalachians in the eastern United States. My recommendation for readers interested in seeing and experiencing these forests would be to take secondary highways and local highways, rather than expressways and interstates in such pursuit. Travel in Scandinavia and in northwestern and northern Russia will yield a similar view of the boreal forests of those regions. Multiple trips through the forested mountains of western North America have caused me to marvel at the altitudinal zonation of the boreal forests that are visible along the highways when crossing any of the mountain systems throughout that region. Trips taken from the urban areas of the Willamette-Puget Sound region to the Olympic Peninsula (temperate rainforest with boreal species), the Coastal Range, and the high Cascades Mountains (boreal forest) provide forest associations to the visitor. The ecotone between the boreal forests and the temperate deciduous forests of eastern North America are quite visible on any travel through regions, such as the northern peninsula of Michigan and the neighboring areas of Wisconsin. Here it is the combination of the Great Lakes coupled with localized physiographic variations that results in a mosaic of stands of boreal and temperate deciduous forest associations in proximity to each other, each representing localized environmental conditions. Travel in any of the forested regions in the eastern United States; in west-central, central, and east-central Europe; in northern China; or in Japan will provide the student with realistic impressions of the temperate deciduous forest communities that we have today. Travel in any forested or brush-covered areas with mediterranean climate conditions, such as those encountered outside of urbanized areas throughout the Mediterranean Basin, in southern California, in south-central Chile, in southern South Africa, and in southwestern Australia, will provide the student with onsite impressions of the plant associations that developed under this climate type. I can only encourage the student of biomes to travel.

I want to take this opportunity to thank several people without whom it would have been impossible for me to undertake the research and writing of this book. My wife Sue Perry has never wavered in her support of this project, both in terms of her active support (taking photographs for me), encouragement, and, yes, even prodding "to get the writing done." I express my particular appreciation to her.

Dr. Susan Woodward has encouraged me as a colleague and fellow scholar to commit to the writing of this book. Her active engagement throughout this process, suggestions, editorial comments, and advice have been invaluable. Radford University has supported me in terms of providing release time, as well as providing material support for travel, printing revisions, and supplying the necessary software for graphic representations. Finally, I also thank the staff at Greenwood Press, and particularly Kevin Downing, for their support in the course of this project. Their assistance has been very valuable to me.

How to Use This Book

The book is arranged with a general introduction to temperate forest biomes and a chapter each on the Boreal Forest Biome, the Temperate Deciduous Forest Biome, and the Mediterranean Woodland and Scrub Biome. The biome chapters begin with a general overview at a global scale and proceed to regional descriptions organized by the continents on which they appear. Each chapter and each regional description can more or less stand on its own, but the reader will find it instructive to investigate the introductory chapter and the introductory sections in the later chapters. More in-depth coverage of topics perhaps not so thoroughly developed in the regional discussions usually appears in the introductions.

The use of Latin or scientific names for species has been kept to a minimum in the text. However, the scientific name of each plant or animal for which a common name is given in a chapter appears in an appendix to that chapter. A glossary at the end of the book gives definitions of selected terms used throughout the volume. The bibliography lists the works consulted by the author and is arranged by biome and the regional expressions of that biome.

All biomes overlap to some degree with others, so you may wish to refer to other books among Greenwood Guides to the Biomes of the World. The volume entitled *Introduction to Biomes* presents simplified descriptions of all the major biomes. It also discusses the major concepts that inform scientists in their study and understanding of biomes and describes and explains, at a global scale, the environmental factors and processes that serve to differentiate the world's biomes.

The Use of Scientific Names

Good reasons exist for knowing the scientific or Latin names of organisms, even if at first they seem strange and cumbersome. Scientific names are agreed on by international committees and, with few exceptions, are used throughout the world. So everyone knows exactly which species or group of species everyone else is talking about. This is not true for common names, which vary from place to place and language to language. Another problem with common names is that in many instances European colonists saw resemblances between new species they encountered in the Americas or elsewhere and those familiar to them at home. So they gave the foreign plant or animal the same name as the Old World species. The common American Robin is a "robin" because it has a red breast like the English or European Robin and not because the two are closely related. In fact, if one checks the scientific names, one finds that the American Robin is *Turdus migratorius* and the English Robin is *Erithacus rubecula.* And they have not merely been put into different genera (*Turdus* versus *Erithacus*) by taxonomists, but into different families. The American Robin is a thrush (family Turdidae) and the English Robin is an Old World flycatcher (family Muscicapidae). Sometimes that matters. Comparing the two birds is really comparing apples to oranges. They are different creatures, a fact masked by their common names.

Scientific names can be secret treasures when it comes to unraveling the puzzles of species distributions. The more different two species are in their taxonomic relationships, the farther apart in time they are from a common ancestor. So two species placed in the same genus are somewhat like two brothers having the same father— they are closely related and of the same generation. Two genera in the same family might be thought of as two cousins—they have the same grandfather, but different fathers. Their common ancestral roots are separated farther by time. The important thing in the study of biomes is that distance measured by time often means distance measured by separation in space as well. It is widely held that new species come about when a population becomes isolated in one way or another from the rest of its kind and adapts to a different environment. The scientific classification into genera, families, orders, and so forth reflects how long ago a population went its separate way in an evolutionary sense and usually points to some past environmental changes that created barriers to the exchange of genes among all members of a species. It hints at the movements of species and both ancient and recent connections or barriers. So if you find two species in the same genus or two genera in the same family that occur on different continents today, this tells you that their "fathers" or "grandfathers" not so long ago lived in close contact, either because the continents were connected by suitable habitat or because some members of the ancestral group were able to overcome a barrier and settle in a new location. The greater the degree of taxonomic separation (for example, different families existing in different geographic areas), the longer the time back to a common ancestor and the longer ago the physical separation of the species. Evolutionary history and Earth history are hidden in a name. Thus, taxonomic classification can be important.

Most readers, of course, won't want or need to consider the deep past. So, as much as possible, Latin names for species do not appear in the text. Only when a common English language name is not available, as often is true for plants and animals from other parts of the world, is the scientific name provided. The names of families and, sometimes, orders appear because they are such strong indicators of long isolation and separate evolution. Scientific names do appear in chapter appendixes. Anyone looking for more information on a particular type of organism is cautioned to use the Latin name in your literature or Internet search to ensure that you are dealing with the correct plant or animal. Anyone comparing the plants and animals of two different biomes or of two different regional expressions of the same biome should likewise consult the list of scientific names to be sure a "robin" in one place is the same as a "robin" in another.

1

Introduction to Temperate Forest Biomes

This volume is restricted in scope to those parts of the world that have what is commonly referred to as a temperate climate. These areas either support now, or have supported in the past several hundred years, a forest cover. Three major world biomes are addressed in this volume: the Boreal Forest Biome, the Temperate Broadleaf Deciduous Forest Biome, and the Mediterranean Woodland and Scrub Biome. These biomes (and particularly the first two) constitute some of the most productive and economically significant forest regions in the world. They represent tremendously large regions covered to a significant extent by a renewable natural resource, trees. A large portion of the world's population has elected to live in these humid temperate climates. The human activities have, for the most part, been ongoing for quite some time and the impacts on the forest vegetation are significant. All three biomes continue to face some kind of alteration as a result of direct and indirect human impacts. These impacts eventually will alter the current geographic boundaries of these forest regions.

An overview of Temperate Forest Biomes in general and their controlling environmental factors is presented in this chapter. The boreal forests of the Northern Hemisphere are covered in Chapter 2. These cold needleleaf forests of the north (which also include significant regions of deciduous trees throughout this biome) form massive forests belts south of the arctic treeline across North America and northern Eurasia. Significant extensions of these forests reach south from these belts in North America, where we find them along the spines of the north-south ranging mountain chains from the Pacific to the Atlantic.

The Temperate Broadleaf Deciduous Forest Biome is discussed in Chapter 3. Its greatest continuous extent is in the eastern half of North America. Another large region occurs throughout northwestern and central Europe from the British Isles to the Ural Mountains; a separate section occurs in the Far East of Eurasia, in China, the Koreas, Siberia, and Japan. Expressions of this biome in the Southern Hemisphere are rather limited in their extent and occur in South America (Chile) as well as in fragmented areas on the southeastern coast of South Africa.

Chapter 4 discusses the Mediterranean Woodland and Scrub Biome. This biome is found in mediterranean climate regions around much of the Mediterranean Sea, as well as in California, southern Chile, South Africa, and southwestern Australia.

The chapter for each biome begins with a general global overview that considers the following aspects:

- Geographic locations of each respective biome
- History of scientific investigation of the biome
- General climatic conditions under which each biome commonly exists
- Types of soils existing and the soil-forming processes that result from interactions between geologic structure, climate, and vegetation
- Vegetation associations of the biome
- Common adaptations and the animals encountered in the biome
- Current conditions and impacts on the respective biome

The general global overview is followed by separate descriptions for each geographic region with an occurrence of the respective biome. Here the reader will find details on the actual locations and their physical environment (climatic conditions, soil types, geomorphologic characteristics, plant associations, and animal species).

Climatic characteristics are a major consideration in understanding the Temperate Forest Biomes. "Temperate" is understood here to involve climates that are warmer than the arctic regions but cooler than the tropics, and that have sufficient moisture to support a forest cover. Climatic characteristics determine to a significant extent the types of vegetation associations that may develop in a particular geographic region, while microclimatic differences often account for the local patchwork of sites with slightly varying species composition within each forest type.

The climate types associated with the three biomes discussed in this volume are summarized in Table 1.1. Climates, however, have undergone significant change during the past 60 million years, and these changes, even though a long time ago, affect the nature of contemporary plant and animal associations. Particularly important are the lasting effects of the Quaternary, especially the Pleistocene Epoch (since 1.6 million years ago [mya]). It had recurring Ice Ages that were characterized by advancing and retreating continental glaciers in the Northern Hemisphere, in both North America and northern Eurasia. Some of the plant associations were in part developed by the middle of the Tertiary Period (20 mya). Climatic characteristics alone, however, are not sufficient to delineate biomes. There are regions, such as in southern Brazil and on the northern half of the South Island of New Zealand, where the climatic characteristics might suggest a temperate broadleaf deciduous forest, but that does not materialize.

Table 1.1 Climate Types of the Koeppen Climate Classification That Pertain to Temperate Forest Biomes

Cf	Mild humid climate with no dry season; all months have sufficient precipitation for vegetation growth with at least 1.2 in (3 cm).
Cfa	Warm temperate climates—mean temperature of coldest month is 64.4° F down to 26.6° F (18° C down to −3° C); sufficient precipitation in all months; warmest month mean is more than 71.6° F (22° C).
Cfb	Warm temperate climates—mean temperature of coldest month is 64.4° F down to 26.6° F (18° C down to −3° C); sufficient precipitation in all months; warmest month mean under 71.6° F (22° C), at least four months have means of more than 56° F (10° C).
Cfc	Warm temperate climates—mean temperature of coldest month is (64.4° F down to 26.6° F (18° C down to −3° C); sufficient precipitation in all months; fewer than four months with means more than 10° C (50° F).
Cs	Mild humid climate with a dry season; precipitation during the driest month is less than 1.2 in (2 cm); 70 percent of the annual precipitation falls during the six months of winter.
Csa	Warm temperate climates—mean temperature of coldest month is 64.4° F down to 26.6° F (18° C down to −3° C); dry season in summer, warmest months mean over 71.6° F (22° C), at least four months have means more than 56° F (10° C).
Csb	Warm temperate climates—mean temperature of coldest month is 64.4° F down to 26.6° F (18° C down to −3° C); dry season in summer; warmest month mean under 71.6° F (22° C), at least four months have means more than 56° F (10° C).
Df	Snowy-forest climate with no dry season; all months have sufficient precipitation for vegetation growth.
Dfb	Snow climates—warmest month mean more than 50° F (10° C); sufficient precipitation in all months; warmest month mean is less than 71.6° F (22° C); at least four months have means more than 50° F (10° C).
Dfc	Snow climates—warmest month mean more than 50° F (10° C); sufficient precipitation in all months; fewer than four months with means more than 50° F (10° C).
Dfd	Snow climates—warmest month mean more than 50° F (10° C); sufficient precipitation in all months; fewer than four months with means more than 50° F (10° C), but coldest monthly mean less than −36.4° F (−38° C).
Dw	Snowy-forest climate with dry winter season.
Dwb	Snow climates—warmest month mean more than 50° F (10° C); dry season in winter; warmest month mean less than 71.6° F (22° C), at least four months have means more than 56° F (10° C).
Dwc	Snow climates—warmest month mean more than 50° F (10° C); dry season in winter; fewer than four months with means more than 50° F (10° C).
Dwd	Snow climates—warmest month mean more than 50° F (10° C); dry season in winter; fewer than four months with means more than 50° F (10° C), but coldest month mean less than −36.4° F (−38° C).

Note: In all C and D climates, temperatures and precipitation are sufficiently high for the growth of large trees. Most of the Boreal Forest Biome is within the D climates; only the southward extensions of this biome in North America reach into C climates. Most of the Temperate Broadleaf Deciduous Forest Biome is in the C climates. However, in North America, some significant extensions into the D climates provide for a large region of mixing between the temperate broadleaf deciduous forest and the boreal forest.

Soil groups associated with the biomes discussed in this volume are limited to a few important ones, most of which are a result of the soil-forming process podzolization (see Figure 1.1). They include the spodosols of the northern forests; the highland soils of the alpine mountain chains and other regions where soils are extremely thin; and the alfisols and their subgroups of temperate humid climate regions (boralfs, udalfs, and xeralfs). These soils have formed largely under current climatic conditions, and they developed with a vegetation cover that is approximately the same as that found today.

The geomorphology of the regions where the biomes of this volume are encountered has a wide range: from the semiflat surfaces that remain after massive continental glaciers scraped the bedrock of old continental cores to the high local relief of geologically recent alpine mountains chains that have been upfolded or uplifted. Almost all types of landforms are found within the Temperate Forest Biomes.

The term "biome" is understood to be a major regional or continental association of plants and animals adapted to the prevailing environmental conditions as the result of endless relationships among the living and nonliving components of the greater ecosystem. Each biome is commonly identified by its climax vegetation type, regardless of the state (for example, stable, transitional, disturbed, and so on) in which the actual plant communities happen to be at the present time. The concept of a climax community comes from observations during the late-nineteenth

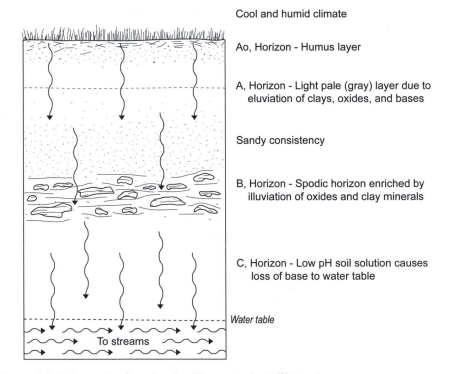

Figure 1.1 Schematic of spodosols. *(Illustration by Jeff Dixon.)*

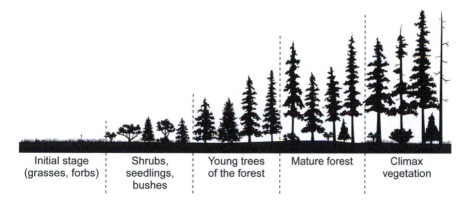

| Initial stage (grasses, forbs) | Shrubs, seedlings, bushes | Young trees of the forest | Mature forest | Climax vegetation |

Figure 1.2 Schematic of ecological succession in boreal forests. *(Illustration by Jeff Dixon.)*

and early-twentieth centuries of the changes occurring in the types of plants occupying particular a site, a process that became known as ecological succession (see Figure 1.2). It is now realized that an orderly replacement of one plant community with another is a great simplification of the complex realities involved in the interplay of all environmental factors in a given stand of trees. Nonetheless, the terminology of climax and successional (preclimax) communities remains useful in describing the dynamics of vegetation and the structural changes undergone largely under the influence of the regional climate.

Ecological Succession

The concept of ecological succession is often used in ecology to describe, in a simplified way, the somewhat predictable changes that occur in a forest environment. The concept of primary succession states that a barren site acquires a plant cover through a series of communities that replace one another. The initial set of plants established on such a site is commonly referred to as the pioneer stage. It often consists of lichens and mosses; grasses and other herbs and shrubs invade later. As subsequent associations of plants become established because of seed dispersal, the early stages of soil formation, and changes in shading and humidity. The pioneer stages will be replaced with a succession of communities. This will continue until eventually the assemblage does not significantly change with time unless major disturbance or climate change reduces the ability of species to regenerate themselves. This final stage represents the equilibrium for that particular site and is commonly referred to as the climax. Members of such a plant association are typically called "climax species." Such a climax association may, if undisturbed, occupy a forest site for hundreds and even thousands of years. If a radical disturbance occurs and clears the site of its vegetation, the process of community development begins again, but it is now known as secondary succession.

Local environmental conditions also may change and favor a different association of plants or at least permit the addition of new species. Succession will commence once again until a new equilibrium between plant associations and the local environment is established and a new and different climax stage is reached.

About Temperate Forests Trees

The dominant growthforms in forests are trees. These are perennial woody plants that have a single main trunk that supports branches well above the ground level. Trees vary in height from dwarf forms to true forest giants, with measured heights of more than 350 ft (105 m). Trees are typically long-lived compared with most other plants. The majority of trees evolved under particular climatic conditions and continue to exist in such environments, with relatively few adaptations to significantly different climates. Trees are geologically young. Most of the presently existing trees emerged during the Tertiary Period or later (66–1.6 mya), although a few might have ancestors as far back as the Triassic Period.

The three Temperate Forest Biomes are characterized by three different types of trees: the boreal forests are dominated by needleleaf coniferous trees; the temperate mid-latitude forests have a dominance of broadleaf deciduous trees; and the mediterranean woodland and scrub species are characterized by small-leaved shrubs that are usually evergreen and frequently have aromatic oils. Of course, other plants are present. However, the processes of adaptation have allowed the characteristic plants to become dominant in their respective biome. The needleleaf coniferous trees are adapted to the northern (boreal) climate with its short growing season, extended winter period with below-freezing temperatures, and often-significant snow loads. Their foliage remains year-round, ready for photosynthesis whenever temperatures increase at the onset of the spring season. The thin needle-like leaves are typically dark green, which allows for a warming of the leaves and subsequent onset of photosynthesis earlier than would be possible if air temperatures alone determined the length of the growing season. Keeping needles year-round also implies that the forest floor in the needleleaf boreal forest is mostly shaded, permitting relatively few shade-tolerant plants to develop under the canopy of a mature boreal forest.

The broadleaf deciduous trees, on the other hand, have adapted to the humid, relatively mild mid-latitude climates with winter seasons that may be typified by periods of freezing to near-freezing temperatures, but with a longer growing season than what is typical farther north. Deciduous trees drop their leaves at the end of the annual growing season and begin the cycle of growth again at the onset of spring. During the leafless season of winter, sunlight reaches the forest floor in this biome, allowing in early spring a warming of the topsoil and the development of the ground-cover and understory plants. These plants will go through their growth stages as the buds of the overstory trees swell and the leaves are just beginning to expand.

The vegetation associations of the Mediterranean Woodland and Scrub Biome have adapted to the climate regime of the mediterranean climate regions: mild winters with hot and often extended dry summers and the rainy season in the winter-half of the year. The characteristic plants of this biome are small-leaved shrubs that are usually evergreen. Their foliage typically has thick waxy coatings that protect the leaves from excessive heat and drying during the drought season of the year. As a protection against being eaten back, this foliage often has aromatic oils not found palatable by foraging herbivores.

Temperate Forest Climates

Comparison of the major characteristics of the three forest biomes (see Table 1.2) reveals that the boreal forests have climates that, in the Koeppen classification, all begin with D. These are the snow climates. The warmest monthly means are more than 50° F (10° C). The coldest months may have extremely low temperatures. In the Dfd climates, the coldest monthly mean can drop to under −36.4° F (−38° C). Most regions have precipitation throughout the year, but some (Dw climates) may have a dry winter season.

The temperate broadleaf deciduous forests have developed in the Cf climates of the Koeppen classification (Cf, Cfa, Cfb, and Cfc). Typically mild humid climates without a dry season, all experience freezing temperatures during the winter season. In all months, precipitation is sufficient for plant growth. A distinctive gradient of temperatures from north to south occurs within these regions, so that the southern areas typically are warmer than the northern areas. In the southern areas, large amounts of precipitation may fall during the summers from convectional storms, Atlantic hurricanes, or Asian monsoons.

The Mediterranean Woodland and Scrub Biome is associated with the Mediterranean or Cs climates of the Koeppen classification (Cs, Csa, and Csb). They all occasionally have frost during the winter season. What sets them apart and what has caused the particular vegetation associations to develop is the distinctive summer dry season when temperatures also reach their highest levels.

Temperate Forest Soils

The climates of the temperate forest regions play a major role in the formation and characteristics of the soils throughout these Temperate Forest Biomes. In the humid boreal and the broadleaf deciduous forests regions, podzolization (leaching of the upper soil horizons and accumulation of materials in lower soil horizons) is the major soil-forming process. Throughout the boreal forests, this typically results in sandy, ash-colored A horizons with accumulations of minerals in the B horizon. In the mid-latitude broadleaf deciduous forests, silicate clay minerals accumulate in the B horizon. The annual fall of leaves and their decomposition enriches the A horizon at least seasonally. With increasing precipitation levels, particularly in the southern parts of these regions, excessive leaching of minerals frequently results in reduced soil fertility.

The soils of the mediterranean woodland and scrub regions are often severely weathered, due to the climatic conditions, and may be thin or stony as a consequence of the long-standing human activities that have resulted in soil erosion over

Table 1.2 Comparison of Boreal Forest, Temperate Broadleaf Deciduous Forest, and Mediterranean Woodland and Scrub Biomes

	BOREAL FORESTS	TEMPERATE BROADLEAF DECIDUOUS	MEDITERRANEAN WOODLAND AND SCRUB
Location	Northern Hemisphere south of arctic treeline	Middle latitudes of eastern North America, western and far-eastern Eurasia, and an isolated small area in the middle latitudes of western South America	Mediterranean Basin and mediterranean regions in New World, South Africa, and Australia
Temperature controls	Subarctic latitude	Mid-latitude seasonality	Mid-latitude seasonality; maritime influences in coastal areas aggravated by long, dry season
Temperature patterns (annual)	Extended cold winters and short mild summers	Well-defined four seasons; freezing temperatures in winter	Mild winter seasons with hot, dry summers
Precipitation controls	Seasonal shift of Polar Front	Summer convectional storms, Atlantic hurricanes, Asian monsoons	Stationary high-pressure cells in summer
Precipitation totals	15–20 in (380–500 mm)	30–50 in (750–1,250 mm)	10–40 in (250–1,000 mm)
Seasonality	Based on temperatures: frozen soils during long winter	Based on temperature: warm to hot summers and cool to cold winters	Based on rainy season in winter and dry summers
Climate type	Cold (boreal) snow climates	Humid temperate climates with distinctive spring, summer, autumn, and cool to cold winter seasons	Mild (mostly frost-free) winter, temperate humid spring, hot-dry summer, and mild, often dry autumn seasons
Dominant growthforms	Needleleaf coniferous trees, usually evergreen	Broadleaf deciduous trees	Small-leaved shrubs, usually evergreen and often with aromatic oils
Dominant soil-forming process	Podzolization	Podzolization	Severe weathering and surface erosion
Major soil order	Spodosols	Alfisols and ultisols	Various, localized xeralfs

Soil characteristics	Sandy ash-colored A horizon; accumulation of minerals in B horizon; generally low in natural fertility	Gray forest soils with accumulated silicate clay minerals in B horizon; some (alfisols) with relatively high natural fertility; more leached soils in southern areas (ultisols) due to higher precipitation levels	Naturally productive soils degraded by thousands of years of human use; rocky subtrates a result of erosion; surface gullies from lack of water infiltration
Typical mammals	Furbearers such as lynx and various weasels; large deer such as moose and elk; all active throughout year	Diverse arboreal and terrestrial forms; some hibernate	Few forms uniquely associated with biome
Biodiversity	Moderate	High	Low to moderate, except in South Africa's fynbos where plant diversity is high
Age	Recent: post-Pleistocene in their current distribution and plant and animal assemblages	Ancient: Tertiary origins	Recent: modern expression of biome may largely be due to human land use practices such as clearing, burning, and grazing
Current status	Climate changes are pushing the northern boundary poleward, while deciduous forests are increasingly invading the biome's southern fringes	Climate changes are altering plant associations; logging, residential development, and transportation corridors fragment forests; reforestation of former agricultural lands are increasing forest area in parts of the United States	Altered by thousands of years of human use, modern impacts from residential development and plant introductions continue to change species composition

vast areas. Small and relatively undisturbed areas do or did have naturally productive soils.

Contemporary Impacts

All three Temperate Forest Biomes suffer from modern environmental impacts. In the case of the boreal and the broadleaf deciduous forests, climate change seems to be at the forefront. The northern boundaries of both biomes are pushing northward with global warming. The northern extent of the boreal forest seems to move northward with increasing average temperatures as deciduous forest plant associations increasingly invade their southern fringes. Both forest biomes suffer from modern logging and mining activities, transportation and other infrastructure development, and, especially in the case of the broadleaf deciduous forests, residential and associated development. The reestablishment of forest trees on former agricultural lands throughout the region of the broadleaf deciduous forests is increasing the total extent of these forests. In the case of the mediterranean woodland and scrub regions, alterations are the results of thousands of years of human use and have changed parts of the biome forever. Contemporary impacts stem from residential development, infrastructure development, plant introductions, and the replacement of the remnant natural vegetation associations with domesticated plants by modern agriculture (for example, the widespread establishment of vineyards throughout mediterranean regions).

Biomes and Global Ecological Zones

The term biome as used in this volume varies significantly from the Food and Agriculture Organization of the United Nations (FAO) highly detailed "Global Ecological Zones," whose purpose was neither a global nor a continental view, but rather a detailed look at individual ecological areas. The descriptions of these zones in combination are good sources of information for the biome as a whole, so FAO zones equivalent to each biome are given in the respective chapter. The knowledge of the characteristics and total spatial extent of each ecological area, as well as the aggregate of the areas in each zone for a particular year, is a baseline that provides the foundation for future assessments of change. Such changes are occurring because of ongoing climatic changes (global warming) and perhaps even more so because of direct human impacts on the respective ecological areas.

Further Readings

Breckle, Sigmar-Walter. 2002. *Walter's Vegetation of the Earth*. New York: Springer.
Chapin, F. Stuart, Mark W. Oswood, Keith van Cleve, Leslie A. Viereck, and David L. Verbyla, eds. 2006. *Alaska's Changing Boreal Forest*. New York: Oxford University Press.

FAO. 2000. Global Forest Resources Assessment. http://www.fao.org/forestry/fo/fra/index.html.

FAO. 2001. Global Ecological Zoning for the Global Forest Resources Assessment. http://www.fao.org/docrep/006/ad652e/ad652e00.htm.

Mittermeier, Russell A., Patricio Robles Gil, Michael Hoffmann, John Pilgrim, Thomas Brooks, Cristina Goettsch Mittermeier, John Lamoreux, and Gustavo A. B. Da Fonseca. 2004. *Hotspots Revisited*. Mexico City: CEMEX.

Woodward, Susan L. 2003. *Biomes of Earth: Terrestrial, Aquatic, and Human-Dominated*. Westport, CT: Greenwood Press.

2

Boreal Forest Biome

Boreal forests are forests that consist largely of cone-bearing needleleaf trees and stretch in nearly continuous belts across northern North America and northern Eurasia. This circumpolar forest belt (see Figure 2.1) coincides over much of the region it covers with what climatologists refer to as the boreal climate zones (boreal meaning "northern"), hence the name "boreal forest." Other terms are used throughout the literature and in translations from other languages or in a direct use of the foreign expression, including evergreen needleleaf forest, coniferous forest and woodland, boreal needleleaf forest, northern coniferous forest, snow forest and taiga (originally a Russian expression), and even tayga, an anglicized version of the last term. The phrases "boreal forest" and "Boreal Forest Biome" are used throughout this chapter to emphasize the fact that the geographic distribution of this forest is primarily in the boreal climate zones of the high latitudes of the Northern Hemisphere. The vegetation association appears monotonously the same on both continents and stretches over millions of square miles on each, with a total estimated size of more than 5 million mi^2 (more than 13 million km^2). This biome accounts for about one-third of all of Earth's total forest area.

Most scientific explorations of the great circumpolar boreal forests belts on both continents have been and largely remain based on economic justifications of one sort or another. Many investigations serve as a scientific foundation for modern forest management to discern the types and growth rates of biomass that are suitable for one type of economic use or another, such as production of saw-timber, pulpwood, fuel, and compressed peat moss. Early on, wood products were of secondary importance to scientific exploration. Forests were recognized as being the

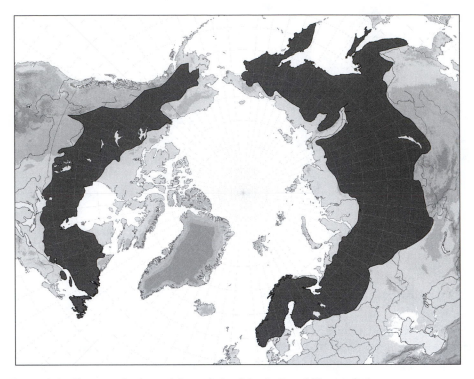

Figure 2.1 Circumpolar boreal forest belt. *(Map by Bernd Kuennecke.)*

necessary habitat for the fur-bearing animals that these early explorers pursued. Indeed, furs were the primary reasons for the early economic exploration of the boreal forests regions of both North America and Eurasia. The hunting, trapping, and trading of furs was the primary reason for the expansion of the Russian Empire across northern Eurasia to the Pacific, and later for the extension of this empire to include Alaska.

In North America, the early explorations of the boreal forests began with the pursuit of fur-bearing animals by hunters, trappers, and merchants who traded with Indians for the pelts. Captain James Cook reached the Pacific Northwest in 1778. His successor, Captain Clerke, had begun trading otter pelts to China a year later. The coastline of the Pacific Northwest of North America was mapped by Captain Robert Gray in 1792. His maps included transportation routes into the interior along the rivers, foremost among which was the Columbia River and its tributaries. The British pioneer Alexander MacKenzie crossed Canada in 1793 and recorded much of the biological wealth that he encountered. Descriptions of the boreal forests of the Rocky Mountains and the regions along the Snake and Columbia rivers, as well as those of the Cascade and Coast ranges were provided by the Lewis and Clark Expedition (1803–1806). The naturalists that worked for the Hudson's Bay Company throughout northern and western Canada made excellent maps of the locations of trees along the Arctic treeline, since wood from trees represented shelter for trappers in winter. In addition, the fuel from wood, as well as its suitability as a raw material for the construction of snowshoes and sleds, made the knowledge of where to find trees rather important to trappers interested in pursuing fur-bearing animals in the tundra to the north of the boreal forests.

Scientific explorations of the boreal forests regions since that time have changed significantly. While descriptions of the natural wealth were important throughout the nineteenth and early twentieth century, the second half of the twentieth century brought about some changes in the focus of such studies. Initially, the correlation between vegetation and climate formed the basis of such investigations. Later, other physical factors such as soils were added to these studies. Scientists such as F. K. Hare attempted to develop correlations between regional climatic characteristics and the boreal forest compositions. Others, such as J. S. Rowe recognized the tremendous complexity of the physical environmental factors of forest stands. He selected criteria to delineate 35 different boreal forest regions across Canada. Leslie Viereck and his colleagues pointed to the important role of fire in the succession of some boreal forest stands, a fact that James Larsen emphasized through his research when he produced seven more generalized delineations of boreal forest categories.

Studies of the boreal forests in Eurasia have been ongoing for a long time, particularly in conjunction with commercial exploitation of timber resources in Scandinavia and western Russia. Long-term studies of the forest ecology of the boreal forests in the Soviet Union began in 1945; at that time, the Serebryanyi Bor Experimental Forest in the Moscow Region was studied.

Boreal forests remain relatively unexplored and our understanding is still somewhat fragmentary. Several efforts to develop models of the ecosystem in order to better understand and comprehend the interdependencies have been undertaken with focus on the Canadian boreal forest regions. Changes in the arctic treeline as an apparent response to global warming influences have been a focus of research efforts since the 1980s by other scientists. Research efforts are often carried out by experimental stations. They concentrate more on understanding the forest dynamics, often in conjunction with developing policies toward sustainable forest practices.

Global Overview of the Biome

The geographic delineation of the Boreal Forest Biomes is characterized by the wide belts in North America and Eurasia (see Figure 2.1), but some subregions are of major significance. These include the mountain vegetation, mountain needleleaf forest, and marine coniferous forest of the western North American Mountain; the extensions along the spines of the higher mountains in Eastern North America to the southern Appalachians; and several of the mountain regions of Eurasia where boreal forests occur at high elevations.

Characteristic for the boreal forest is a relatively low number of tree species growing in a mosaic of plant communities throughout the distribution regions. The composition of these plant communities is determined in general by the physical environment (climate, geology, geomorphology) of the region, as well as by local soil and drainage conditions. Also of major importance is the influence of the fire history of each subregion or area. Most of the regions supporting boreal forest biomes had been glaciated during the Pleistocene glaciations. They are covered by irregular ground moraines and other glacial features and often have poorly developed drainage systems. As a result, bogs and other wetlands are commonly encountered throughout the boreal forests of both continents. A patchiness of permafrost throughout the northern sections of this biome further contributes to a quickly changing mosaic of successional, subclimax, and climax plant communities, all of which are highly sensitive to changing environmental conditions.

In general, four genera of conifers dominate on both continents: spruces, firs, pines, and larches (see Figure 2.2). The occurrence of these conifers as the dominant trees has contributed significantly to the popular use of the name Northern (Boreal) Needleleaf Forests. Two genera of broadleaf trees are commonly encountered, however, and are indeed indicative for much of the region in areas of recent or frequent disturbance: birches and aspens or poplars. These deciduous trees grow

Figure 2.2 Spruce-pine-birch boreal forest. *(© L. Sue Perry, by permission.)*

Telling Fir from Spruce

Firs (*Abies*) and spruces (*Picea*) are both coniferous evergreen trees in the Pinaceae family. Both can be large trees reaching heights of more than 200 ft (65 m). One distinguishing factor of firs is that the needles are attached to the branches by a small base that has the appearance of a miniature suction cup. A second and more visible characteristic is that the cylindrical seed cones of firs grow erect on the branches.

Spruces, on the other hand, typically have whorled branches (pointed slightly upward and growing in a spiral fashion out of the trunks). Their needles are also attached to the branches in a spiral form. Their seed cones hang downward from the branches, often at an angle (see Figure 2.3).

Figure 2.3 Red spruce. *(© Susan L. Woodward, used by permission.)*

along waterways and near the edges of ponds and lakes. Different species of birch and aspen frequently occur in different geographic segments of the biome.

Boreal forest largely overlies formerly glaciated regions and areas of discontinuous or patchy permafrost on both continents. The northern boundary of the Boreal Forest Biome corresponds to the southern arctic treeline: north of this line trees generally do not grow, probably as a result of permafrost in the upper surface layers. Here the Boreal Forest Biome meets the Tundra Biome. On the North American continent, this treeline angles across the continent from west to east. Beginning at about 60° N at the northwestern coast of Alaska, it skirts along this coastline northward to about 68° N, follows this parallel eastward to 120° W, slants downward and eastward to nearly the southern shores of the Hudson Bay, before arching northeastward direction across Labrador and meeting the Atlantic Ocean at 58° W in eastern Canada. In Eurasia, the northern treeline begins on the northwestern coast of northern Norway and continues eastward as a continuous line a hundred miles to the south, only diverging southward to any major extent east of the Ural Mountains in central Siberia. It appears that on both North America and Eurasia the growth of trees northward of this line is prevented by the continuous layer of permafrost (that is, permanently frozen ground) close to the surface. Global warming will push the arctic treeline northward. The extent of that future movement remains to be seen.

The core segments of the Boreal Forest Biome exist in the form of broad forest belts that have an average width of 600 mi (1,000 km) across much of North America, although the width can measure as much as 1,250 mi (2,000 km)—the distance from the Mackenzie River delta to the mountains of southern Alberta and northern Montana. In western Eurasia (Scandinavia and western Russia), the belt is similarly wide (approximately 600 mi or 1,000 km). East of the Ural Mountains, in Eurasia, the boreal forest belt extends about 1,850 mi (3,000 km) toward the Pacific Ocean. It also broadens in its north-south extent. It reaches from near the Arctic coast to south of Lake Baikal, and as it extends eastward it reaches south to the northern boundaries of Mongolia and China. On both continents, some southward extensions of the boreal forests and isolated geographic regions in the form of isolated mountain blocks with boreal forest-type cover occur. On the North American continent (see Figure 2.4), where the boreal forest covers nearly 28 percent of the land area, several southward extensions follow mountains out of the northern belt of boreal forests. Rising elevation dictates a cooler climate until the upper limit of the forest is reached at treeline. One extension is along the Coastal Ranges, stretching from northern Canada into northern California. In southern British Columbia the Cascade–Sierra Nevada range breaks off from and extends southward of the coastal mountains a little to the east of the Coastal Ranges, leaving the Puget Sound–Willamette Valley–Rogue Valley to the west and connecting with the Coastal Range once again in the Klamath Mountains of southern Oregon and northern California. From there, the Sierra Nevada, continues east of the Great Valley southward toward the Mexican border. A second and, in parts, rather broad extension stretches along the Rocky Mountains from northern Canada to near the southern boundary of the United States. More so than in mountainous regions farther west, the Rocky Mountains have significant areas of high elevation that are above the treeline and form virtual islands of alpine tundra in a sea of boreal forest that stretches along the backbone of this massive mountain chain.

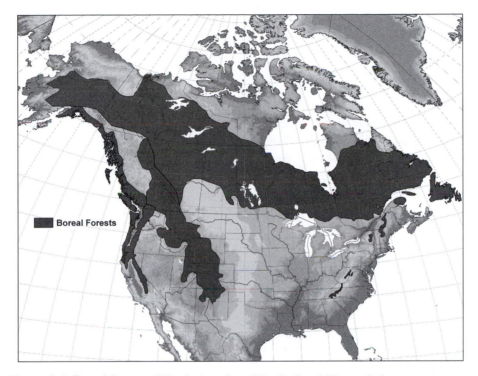

Figure 2.4 Boreal forests of North America. *(Map by Bernd Kuennecke.)*

Some large "outlier" regions with boreal forests occur in form of mountain blocks that, due to their elevations and exposure to moisture (orographic effect), have climatic characteristics conducive to the development of this forest type. Examples include the Blue Mountains and the Wallowa Mountains of Oregon. A third extension of the boreal forests extends southward along the northern Appalachian Mountains. This region, in particular, has a number of transition areas (ecotones) in which the boundaries between the Boreal Forest Biome and the Temperate Forest Biome are fluid.

Climate

Boreal Forest Biomes, by their very name, imply cool to cold (but *not* arctic) climatic conditions that prevail throughout much of the year. The broad belts across North America and Eurasia coincide with the climate called variously Cold-temperate Boreal Zone, Microthermal Climates, or Snowy-forest (microthermal) climates. In the most commonly used climatic classification, the Koeppen climate system, the Boreal Forest Biome corresponds with the subarctic and the cold continental climate types (Dfc, Dfd, Dwc, and Dwd) (see Figures 2.5, 2.6, and 2.7; see also Table 1.1).

Orographic and Rainshadow Effects

Whenever a mountain range is in the path of prevailing winds, then both orographic and rainshadow effects commonly take place.

An orographic effect occurs when warm moist air is pushed with the prevailing winds against a mountain range. As the rising air cools, water vapor is condensed and falls out as precipitation on the side of the mountain that is exposed to the winds. Clouds are common on the windward sides of mountains. These airmasses, depleted of their moisture content, continue to flow across the mountain range, and as dry air descends the other side, it warms rapidly. Dry, warm air continues to flow in the prevailing wind direction across the downwind or leeward side of the mountain range, which is commonly said to be in the "rainshadow" of that range.

An example will help illustrate these phenomena. Eugene, Oregon, is on the windward side of the Cascade Range, and, because of orographic effect, receives about 47 in (1,185 mm) of rainfall per year. Bend, Oregon, on the eastern side and in the rainshadow of the Cascade Range, receives about 12 in (296 mm) of rainfall a year. If one travels between these two cities across the McKenzie Pass (5,325 ft above sea level) and stops at the summit, one can actually see the difference: looking west one overlooks the lush Grand fir and Douglas fir forests that have developed under conditions of heavy annual precipitation; to the east, one overlooks forests of widely spaced ponderosa pine; and still farther to the east, one can see the high desert with its sagebrush and bunchgrass cover.

Within these climate regions, we currently find up to six months with average temperature at or below freezing. Summers are typically short and cool to warm. There may be anywhere from 50 to 100 frost-free days. The coldest month average temperature is under 26.6° F (−3° C). The average temperature of the warmest month is above 50° F (10° C). The 50° F summer isotherm marks the approximate poleward limit of forest growth. Extensions of the Boreal Forest Biome reach into other climate regions, such as the marine west-coast and mediterranean (Cfc, Cfb, Csb) climates along the west coast of North America, the H climate region in the mountainous interior of North America, and even the humid subtropical climate (Cfa) along the mountain ridges in the eastern part of North America (see Table 1.1). One of the distinguishing climate characteristics of the boreal forests is the great range in temperatures experienced in a year. Exceptions to this occur only in cases in which proximity to the ocean has a moderating influence upon high temperature fluctuations. The greatest range in mean monthly temperatures within the boreal forest region occurs in eastern Siberia. In Yakutsk, Russia, temperatures in January average below −40° F (−40° C) while July temperatures average above 61° F (16° C). In Verkhoyansk, Russia, recorded extremes of −90° F (−68° C) and 90° F (32° C) have been recorded. In North America, the maximum range of mean monthly temperatures (110° F or 44° C) occurs in the interior of Alaska.

Precipitation is of critical importance to the development and growth of forests. Mean annual precipitation throughout the wide belts of the boreal forests in North America and Eurasia is between 15 and 20 in (380 and 500 mm) per year. While such precipitation levels would be considered low in warmer humid climate zones, the generally low temperatures in boreal forest lands cause low evaporation rates and therefore sufficient moisture is available to support the growth and development of forests. Furthermore, countless lakes, ponds, rivers, bogs, marshes, fens, and other wetlands interspersed with the forests hold

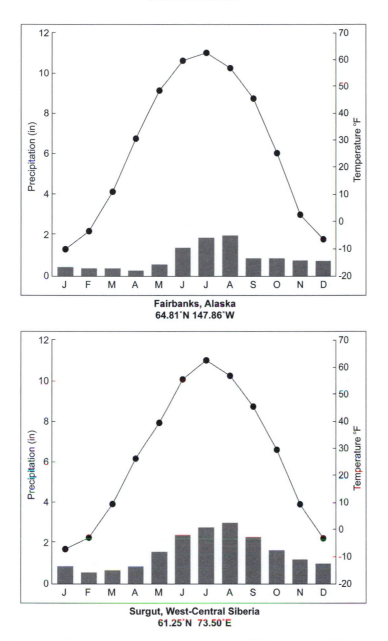

Figure 2.5 Examples of climographs for Dfc climates. *(Illustration by Jeff Dixon.)*

tremendous volumes of water. Winters are mostly dry (winter drought is extreme in eastern Siberia); more than half the yearly precipitation occurs in summer.

In North America, the northern limits of the biome coincide with the summer position of the Arctic Front. This is a meeting place of different airmasses. To the north, dry arctic and continental polar airmasses dominate the climate. The

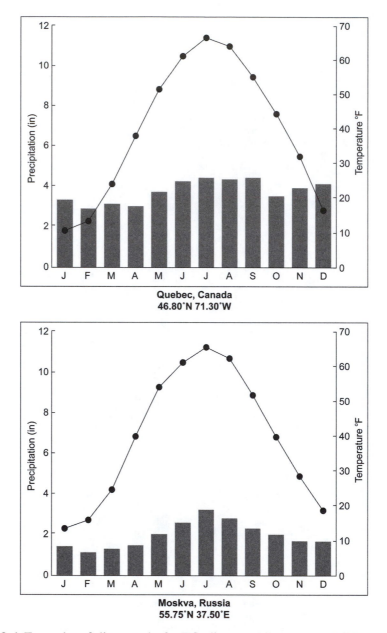

Figure 2.6 Examples of climographs for Dfb climates. *(Illustration by Jeff Dixon.)*

southern boundary of the biome corresponds with the southernmost position of the arctic front, and in winter the line where the annual minimum temperature averages −40° F (−40° C). A corresponding relationship between minimum temperature and the southern limit of the Boreal Forest Biome has not yet been documented for Eurasia. The climatic conditions of the "extensions" and "islands" of the Boreal Forest Biome are somewhat more complicated.

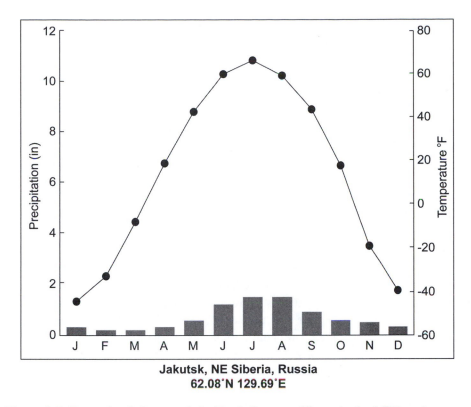

Jakutsk, NE Siberia, Russia
62.08°N 129.69°E

Figure 2.7 Example of climograph for Dwd climates. *(Illustration by Jeff Dixon.)*

Geology

The underlying geologic composition and structure of the major areas of the Boreal Forest Biome, for example, the wide belts in both North America and northern Eurasia, are varied. The Laurasian shields (the Canadian Shield of North America, the Baltic Shield of Norway, Finland, and the northwestern part of Russia, and the Angara Shield of northeastern Russia in Siberia), which are parts of Earth's oldest continental shields, consist mostly of crystalline rock ranging in age from more than 2 billion to at least 570 million years (all are Precambrian). The sedimentary covers of the sections adjoining the old continental shields are found in North America between the Canadian Shield and the Rocky Mountains, as well as in a smaller region along the southern shores of the Hudson Bay. In Eurasia, *sedimentary covers* underlie forests south and east of the Laurasian Shield in the west and extend from western Russia to the Angara Shield in Siberia. It is interrupted only by the Ural Mountains. In the western regions of the biome in North America, the alpine system is a massive structural region encompassing the Rocky Mountains and their northwestern extensions, the Alaska Range and the Brooks Range. In Eurasia, an alpine structural region is encountered in the eastern geographic regions of the biome in northeastern and eastern Eurasia—an extension of the alpine system from northwestern North America—and a part of the world-girdling system of mountain chains that formed since the late Mesozoic Era.

Geomorphologic Characteristics

All parts of the Boreal Forest Biome have been subjected to the long-lasting influences of the Pleistocene glaciations. For most geographic regions, this involved the movement of tremendously large and enormously thick sheets of ice across the surface for a period of several hundred thousand years. Every time the ice moved, its weight pressed down on the underlying rock materials, gouging, scraping, and polishing the bedrock and moving the generated debris along with it. This debris assisted in the scouring action upon the bedrock. The advancing ice had a flattening and smoothing effect on the land over which it moved. The ice sheets apparently underwent several stages of advancement and retreat. When the ice retreated (that is, melted at its forward edge), it left behind the transported debris. Deposits of various forms, including moraines, eskers, drumlins, kettle holes, and outwash plains are some of the resulting landforms. Debris left by ice was often in such large volumes that meltwater could not always break through the dams that it created. Innumerable lakes of various sizes formed throughout the regions now are covered by boreal forests, and the surfaces left after continental glaciation often still have poor drainage due to the irregular depositional topographic features. The relative flatness of terrain sculpted by continental ice sheets is today characterized by wide drainage systems with streams flowing at relatively slow speeds.

In high alpine regions, the Pleistocene ice collected, often in previously existing stream valleys, and gouged U-shaped glacial troughs. Where alpine (or valley) glaciers terminated, they deposited large volumes of debris, and often dammed the meltwater stream with glacial end moraines, creating many high mountain lakes. Many such lakes remain.

Soils

The soils are thin in the geographic regions recently scoured by Pleistocene ice sheets. Current precipitation and potential evapotranspiration are both low in the northern boreal forest belts. Thus, soils are generally moist and may be either partially or totally frozen beneath the surface in the northern latitudes close to the arctic treeline. When this occurs, permafrost forms a discontinuous layer.

The end of the Pleistocene Ice Ages occurred only a relatively short time ago, geologically speaking, and the soil development is recent and possibly incomplete, as indicated by the shallowness of some of these soils. This variability of soil developments is reflected in the current distribution of plant communities. Some favor dry sites, while others prefer wetter locations; some plants are able to tolerate low nutrient conditions, while others cannot do so.

According to the U.S. Soil Taxonomy, the predominant soils are spodosols, although the general term for podzols (derived from the Russian word for ash) is often used. The soil-forming process of podzolization occurs as a result of the acid soil solution that forms from the slow decomposition of organic matter under needleleaf trees. When such acid soil solutions travel downward, iron and magnesium are leached from the upper soil horizon (the A horizon) and accumulate or are deposited in the B horizon. As moisture filters through the upper soil layers, clays and other minerals are moved downward to lower layers. This leaves the upper layer with a relatively sandy texture (see Figure 1.1). The limited activities of microorganisms in the soil layers (especially the upper ones) allow a dense mat of needles to form on the soil surface. The decomposition process, nonactive when the soils are frozen, is further slowed by the highly lignified nature of the needles. The slow leaching of acids (predominantly tannin) from this mat into the upper soil layer, in conjunction with the shading effect from the evergreen needleleaf tree cover that keeps evaporation to a minimum, results in wet and often acidic upper soil layers. Poor drainage and waterlogged soils keep nutrient recycling at a minimum (compared with more southern forest biomes) and significantly affect the vegetation mosaic and the vegetation cycles of this biome.

Vegetation

The foremost characteristics of the Boreal Forest Biome are derived from the major plants of this biome: coniferous trees (often referred to as gymnosperms) that have evolved to survive and even thrive under the generally demanding climatic, as well as to the challenging conditions presented by the mostly thin acidic soils of these regions. Despite this dominance in growthform, however, relatively few species in only four genera occur: the evergreen spruces, pines, and firs, and the deciduous larches or tamarack.

The vegetation structure of the boreal forest is typified by a single tree or canopy layer (see Figure 2.8). On both continents, this layer consists of two or frequently only one kind of tree. Throughout the North American section of the biome, on some sites the dominant trees are one or two kinds of fir. In the Eurasian boreal forest belt, in Scandinavia and western Russia, Scots pine is common and often the dominant tree of the taiga. Spruces here are Norway spruce and Siberian spruce. Throughout the eastern part of this belt, in the vast regions of Siberia and in wet areas, larches dominate, especially along the northern fringes of the forest belt. Below the canopy layer of trees, few if any shrubs grow. The lowest or ground layer often consists of a groundcover made up of lichens or mosses. This layer may also contain some low shrubs, such as members of the heath family, and the occasional widely spaced herb, such as wood sorrel. Throughout both continents, broadleaf deciduous trees and shrubs are common in the early successional communities. This holds true for both primary as well as for secondary succession. The most common broadleaf deciduous trees on both continents in

Figure 2.8 Vegetation profile of the boreal forest. *(Illustration by Jeff Dixon.)*

natural vegetation successions are alders, birches, and aspens. A cyclic succession apparently happens that involves the nitrogen-depleting spruce-fir forest and the nitrogen-enriching and accumulating aspen forests alternating on the same site over time. Estimates concerning the rotation cycles of the vegetation climax communities of the Boreal Forest Biome on both continents place the cycle at 200-plus years.

The characteristic evergreen conifers (firs, spruces, and pines) as well as the deciduous larches of boreal forests are typically conical or spire shaped. They have adapted to the prevailing cool to cold temperatures, the long and hard winter season, the physiological drought that typically occurs during the winter, and the short growing season in a number of ways. The conical shape of the trees seems to help them, at least partially, to shed snow and thus reduce the incidence of broken branches resulting from snow weight. The needleleaf shape of their leaves is an advantage since the relatively thin narrow shape represents a reduction in the surface area through which water may be lost during transpiration. This is of special importance during the winter season, when drought conditions are often present and when the freezing of the ground prevents trees from replenishing moisture from soil water. In addition, the needles of the boreal conifers have a rather thick waxy coating (cuticle) that is waterproof. The stomata are deeply sunken into this waxy layer, which further protects the needles from the drying effect of moving air-masses, especially the dry winds that accompany the drought conditions existing during much of the winter season.

A short annual growing season, combined with rather severe winters through-out the boreal forests, favors the evergreen firs, spruces, and pines. Retention of the needles permits these trees to begin the process of photosynthesis as soon as winter breaks and temperatures begin to climb in spring. This gives evergreen conifers an advantage over deciduous broadleaf trees: they do not have to waste precious time during the short growing season simply to grow new leaves. The dark green color of the needles further assists these trees in the absorption of maximum solar energy to increase leaf temperatures and to restart the photosynthesis process as early in the year as possible, even on an intermittent basis when temperatures still fluctuate significantly in early spring.

An exception to the evergreen conifers are the larches, which are often called tamarack in North America, especially in the Northeast and northern Midwest. Larches are deciduous. They are a colorful addition to the dark evergreen coni-fers: their needles are typically somewhat lighter in color, tending to chartreuse in spring and summer. It is in fall, however, that the differences between the dark evergreen conifers and the larches really come to shine: the larch needles turn golden before they are dropped, providing a bright color to the otherwise monoto-nous, dark evergreens of boreal forests. The deciduous larches are commonly dominant trees in those areas of the boreal forest that have an underlying layer of nearly continuous permafrost. Larches have developed and persisted in regions where the climate is too dry and too cold for even the waxy needles of the spru-ces, firs, and pines.

Very long winter periods imply not only bitterly cold temperatures, but also the lack of water while it is frozen in the soil layers. Leaves of trees that are exposed to the winter winds are highly susceptible to severe damage from water loss that results from evaporation. Several properties of the needles of boreal species act to minimize such loss. The trees of the boreal forest are essentially dormant for the duration of the dark and typically cold winter. When temperatures begin to drop significantly in autumn, the foliage of the conifers seems to undergo a process of "hardening" during the late autumn period. This increases the resistance of the foli-age to frost and thus prevents damage to the needles when temperatures drop as low as of $-76°$ F ($-60°$ C). This hardening process is a matter of survival for many trees. For example, spruce needles would be killed at $19°$ F ($-7°$ C) if they did not previously harden when temperatures fell at the onset of fall. In some of the interior regions of the boreal forest in east-central Siberia, winters can become exception-ally cold and dry. Under such circumstances, even the properties of leaf hardening are insufficient to permit survival. As a result, the deciduous habit of larches and birches is an advantage. Shedding of the foliage at the onset of the cold season when temperatures drop below what freezing foliage can resist becomes a necessity to prevent the destruction of the life-giving photosynthetic surfaces and the killing of the trees. This is the singular most important reason why deciduous larches and broadleaf birches and aspens prevail in the coldest regions of the boreal forest.

Throughout most of the Boreal Forest Biome, the height of the trees is in general not very tall (50–80 ft or 15–24 m). Nearing the arctic treeline, stature decreases significantly until dwarfed conifers and larches mix with the plants of the tundra ecotone. Some significant exceptions to the often small stature of the trees in the boreal forest belt of North America occur in the southward extensions of the biome in the Canadian and the U.S. Pacific Northwest, in particular, where some of the world's tree giants, such as Douglas fir, grand fir, and western hemlock grow. In the California portions of the southward extensions of the boreal forest, the coastal redwoods and the giant sequoia can reach heights of more than 200 ft (65 m).

Climate is commonly zoned according to latitude (especially in the absence of high mountains), because temperatures drop with increasing distance from the Equator. Most of our broad boreal forest belts across the northern parts of North America and Eurasia lack high mountains, so latitudinal zonation is strong. Toward the northern limits of the boreal forest, an ecotone, or mixing zone, with the tundra biome occurs. Boundaries between the two biomes are rather fluid, and the boreal forest often penetrates deep into this zone. Throughout the forest-tundra ecotone, conifers are often scattered (microclimatic conditions permitting their growth make this possible), and mostly they are dwarfed by harsh climatic conditions at the northern boundary of the biome. Only those parts of the conifers that are covered by snow and thus insulated from the fiercely cold winter winds and their freeze-drying effect can survive the winters without too much damage. These scattered outliers of conifers from the boreal forest often grow under environmental conditions that retard or even prevent normal seed germination. One adaptation to such climatic limitation is to reproduce by a process called layering. Dwarfed deciduous broadleaf trees also occur at the northern edge of the boreal forest. They survive at or near the northern treeline because their short stature and deciduous habit adapt them to the extreme environment.

South from the forest-tundra ecotone into the broad belts of boreal forests across North America and northern Eurasia, open-canopied stands of mostly coniferous forest make up the next latitudinal zone. They typically have a continuous groundcover of dwarfed shrubs and a nearly uninterrupted surface layer of mosses and lichens. This somewhat-open forest immediately to the south

···

Layering (Or Ground Layering)

In climates where conditions are too severe to allow plants' normal development of flowers and seeds, such as in the high latitudes where cold temperatures may prevent seed production, plant propagation often occurs through the process of layering. This is a rather slow process that involves the sprouting of roots from branches that touch the ground (that is, the development of adventitious roots). From the point of contact, a new leader branch will grow upward and subsequently form a new individual tree. The new roots support the growth of a full plant from such a branch. If the branch becomes separated from its parent tree, a separate tree exists. Layering is a type of cloning or asexual reproduction.

···

of the boreal-tundra ecotone is referred to by most scientists as the actual taiga (see Figure 2.9).

In North America, the next zone to the south is the broad belt of the needleleaf evergreen forests that so characterizes the biome with its closed canopy of spruce and fir trees. The southernmost zone of the boreal forest is, once again, an ecotone or mixing zone—actually several mixing zones exist that are slightly different from one another in terms of their plant composition. From west to east, different vegetation types, corresponding to the respective climatic conditions that characterize North America south of the boreal forest from the western mountains to the Atlantic coast, interfinger with the boreal forest. East of the slopes of the western mountains, the boreal forests first mix with the plant communities of the Temperate Grassland Biome and its short grass and tall grass prairies. This ecotone extends eastward to the Great Lakes. The manifestation of this mixture of species from the boreal forest and the grasslands usually occurs in the form of aspen parklands, open areas covered by short and long grasses with scattered aspen and even groves of aspen. From the Great Lakes eastward, environmental conditions change to a temperate humid climate and permit the growth of broadleaf deciduous trees. As a result, the ecotone between the Boreal Forest Biome and the Temperate Broadleaf

Figure 2.9 Russian taiga. *(© Susan L. Woodward, used by permission.)*

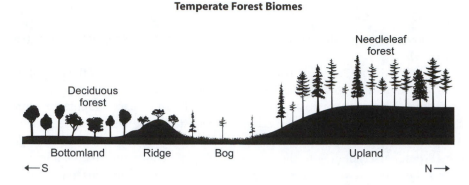

Figure 2.10 Transect through boreal forest. *(Illustration by Jeff Dixon.)*

Deciduous Forest Biome is characterized by a mixed forest of needleleaf evergreen trees such as the eastern white pine and eastern hemlock, and the broadleaf deciduous trees such as northern red oak, American beech, and sugar maple. Some scientists include this ecotone as an integral part of the Temperate Broadleaf Deciduous Forest Biome, rather than a mixing zone between the boreal and the temperate broadleaf deciduous forest (see Figure 2.10).

On the Eurasian continent, the northern ecotone of the boreal forest is a replica image of that in North America. However, the climatic conditions along the southern fringes of the boreal forest are different in Eurasia. It is somewhat drier and the increasingly warmer conditions there at the southern fringe of the biome imply an increase in evapotranspiration that is not compensated for with higher rainfall levels, as found in the eastern half of the southern fringes of the boreal forest in North America. The southern ecotone of the biome in Eurasia thus is primarily a forest-steppe transition in which the grasslands gain dominance with decreasing latitude.

The Vegetation Mosaic of the Boreal Forests

The Boreal Forest Biome is typified by a mosaic of plant communities that represent particular stages in the successional process. Localized factors such as climatic conditions, surface configuration and drainage, nutrient availability and levels, fire and its history, stage within the plant succession, and seed availability play an enormous role in the patterns of this vegetation mosaic. In mature stands, some conifers of the boreal forest prefer the drier and better-drained sites. Other conifers seem to have more of a preference for the moister sites, which are often poorly drained. To a large degree, the vegetation mosaic is a frequently repeated pattern of wetlands and forests.

Muskeg. Low-lying, water-filled depressions typically occur in poorly drained former glacial depressions. The most distinctive of the boreal wetlands is the northern bog or muskeg. The term was coined by Algonquian Indians and means "trembling

Earth." As one walks across the surface, ripples radiate outward from each step. (The feeling that one gets when walking across a drier section of a bog—where it does support the weight of a human walking across its surface—is as if one is walking on a trampoline.) In these wetlands, sphagnum moss, or the common peat moss (*Sphagnum* spp.), forms thick spongy mats that float on a body of standing water, at least throughout the wet spring season. This mat, in turn, becomes the growing medium for plants more common to the Tundra Biome, such as small shrubs of the heath family—for example, Labrador tea, bilberries, and cranberries—and some herbaceous plants. The bog rose is a terrestrial orchid commonly associated with these bogs. Sedges that are often referred to as cotton grasses are common indicators of muskeg. Typically larch and black spruce develop on the edges of such ponds and may indeed encroach upon the margins in a dwarfed form. The bog environment is highly acidic (pH = 4.0). As a result of the moss mat and the fact that the water is mostly standing water, little interchange with freshwater streams occurs; thus muskegs are low in nutrients. Some highly unusual plants have adapted to these acidic nutrient-poor conditions, including the carnivorous plants that trap and digest insects to provide themselves with a source of nitrogen. Among these we find plants like the pitcher plant and the sundews.

The typical surface of bogs or muskegs consisting of a thick sphagnum moss cover deserves some additional attention. Sphagnum or peat moss is an interesting material that has been found to have a significant effect on water-logging, and thus affects the subsequent development of plants in the area of its growth. Scientists have found that, once sphagnum moss is established, it has the capacity to hold up to 40 times its dry weight in water. With such a tremendous amount of water held, sphagnum moss can indeed limit the types of plants that are able to grow in such an environment, since the roots of most plants need at least short periods of dryness. When sphagnum areas develop, they cause the release of moisture through evaporation and thus affect the microclimatic conditions downwind, helping establish sphagnum mats in the immediately neighboring areas as well.

Most bogs or muskegs developed as steps in the long-term development from the gouged-out glacial ponds of the Pleistocene Ice Ages to dry land (see Figure 2.11; see also Figure 2.10). In addition, some of the muskegs develop as a result of paludification of spruce forests (see below). Scientists believe that almost one-third of western Siberia is covered with muskegs and similar wetlands. The water-logging of vast areas is an annual occurrence and perpetuates the existence of these bogs. Spring arrives first in the southern portions of the watersheds, and the meltwater flows north in this region, following the slope to the Arctic Ocean. Since the northern portions of this region are still frozen, this meltwater is blocked by ice on the north-flowing rivers, causing extensive flooding of the alluvial floodplains adjoining these rivers. Drainage is poor in the nearly level floodplains, so ponding occurs and sets the stage for bog development.

Slightly different from bogs is the situation of marshes and swamps. Marshes that are mostly occupied by sedges are often referred to as fens. (This is also a term

Figure 2.11 Overgrown muskeg. *(© L. Sue Perry, by permission.)*

commonly used in Britain for coastal marshes with brackish water). Swamps (wooded wetlands) occur only where water is flowing (even if flowing very slowly). In some cases, short-lived swamps are created when beavers dam streams. A rapid change in vegetation types occurs in such affected areas, because willows, poplar, and alder typically thrive under such conditions, while less water-tolerant species are literally "drowned-out." The vegetation that develops provides cover and is a food source for a number of mammals and birds.

The forest patches. The pine forests that occur in the Boreal Forest Biome in North America are dominated by the Jack pine. This dominance is restricted to those areas where relatively dry soils occur, such as on former dune areas and on sandy outwash that developed during the melting of continental glaciers at the end of the Pleistocene Ice Ages. These sandy areas are extremely well drained and therefore relatively dry, and the soils are so low in nutrients that neither fir nor spruce can germinate and grow.

Almost pure stands of larch typically are encountered in those areas that are level and underlain by permafrost. These areas often exhibit waterlogged soils, since water can neither drain downward nor easily move laterally due to the lack of significant slope. Larches have a greater tolerance to waterlogged substrate than pines, spruces, and firs. Such larch forests often have an open canopy with an understory consisting of low shrubs and a groundcover of mosses and lichens (in some areas as a continuous groundcover layer). Pure larch forests are highly localized in parts of Alaska and are not a common phenomena. In Eurasia, however, pure larch forests are more frequent, particularly in Siberia east of the Yenesei River. The extreme continental climate conditions allow the persistence of a nearly continuous permafrost layer across vast areas in this region and provide the environmental conditions under which larches attain dominance.

Vegetation Cycles in the Boreal Forests

The boreal forest is in a state of constant change (even though this change may take place over long periods of time). The classic textbook ecological succession that is described for temperate forests in all likelihood does not occur in the Boreal Forest Biome. Instead, all indications are that the forest mosaic is never stable for long, but shifts frequently. Change is triggered as a response to soil moisture, fire, depth to the underlying permafrost, thickness and composition of the organic layer on the forest floor, and nutrient depletion, which often occur with changes in the makeup of the organic soil layer. The interconnections of these complex processes are not yet fully investigated and explained. However, some of the likely ways in which one plant community is superseded and replaced by another in these ongoing cycles of forest growth and gradual decline are identified below.

By all accounts, fire is the most important factor in this continual change. Fires in the boreal forest are relatively common (lightning being the most frequent cause); and fires reoccur at intervals of 50 to 200 years in much of the North American boreal forest, although intervals of up to 500 years seem to be more common in the moister eastern sections of the biome. Not every forest fire is large; indeed, the average fire ranges over relatively small areas of about 10–12 ac (4–5 ha). However, the Boreal Forest Biome is known to have had some of the largest forest fires in the world, extending to well over 250,000 ac (100,000 ha) and causing a tremendous amount of damage over a wide region.

In North America the Jack pine and in northern Eurasia the Scots pine are among the few trees that, because of their sufficiently thick bark, can tolerate fires as adult trees. Such a tolerance is found only as long as the burn temperatures are average or low, the fires are confined mostly to the forest floor, and the time of exposure to fire is relatively short. When fires have been suppressed for long periods of time and, as a result, fuel (burnable biomass) has accumulated on the forest floor, very hot fires can occur. Adult trees that have not been pruned of dead branches by high snow loads or high winds can contribute to hot crown fires, which are not infrequent in these environments and helped along by periodic droughts. When hot crown fires happen, even the normally fire-resistant adult trees will be killed. Hot fires frequently kill trees over large areas. A regeneration of the whole stand will occur in such cases, complete with the natural successional stages and their respective dominant vegetation. Some tree species of the boreal forest have adapted to repeated fire events. Jack pine and black spruce maintain closed (serotinous) cones on their branches, and their cones open only when heated by the temperatures of wildfires. Their seeds will fall to the ground where the fire has just removed competitive vegetation. Thus, these trees replace themselves rather rapidly after a burn. Other common pioneers that invade and become established after a forest fire include birches, aspens, and poplars—broadleaf trees with light seeds that are easily dispersed by even gentle breezes. They can become established over vast recently burned areas and are a significant part of the initial forest succession on such sites newly opened by fire. If the aspens or other deciduous trees should be killed by rapid but hot surface and crown fires, they may actually resprout from their roots, or if a section of bark at the stump was protected from the heat of the fire, they can resprout from the stumps. In this way, if fires occur with sufficient frequency, broadleaf trees may regenerate themselves and keep conifers from becoming established and replacing them. The results of fires are frequently even-age stands of a single species that replace the needleleaf forest. Since fires are somewhat spotty in their distribution and are not always extremely large, the subsequent early succession with even-age, single-species stands result in a patchwork quilt of different tree species in stands of different ages.

When no fires initiate the succession of forest stands in North America (for example, on well-drained sites), we find that the aspen and birch stands will, over time, be invaded by the more slowly dispersing white spruce. After 80 to 200 years (mostly after about 150 years), the relatively short-lived birch, aspen, or poplar will have died off.

Their seedlings have high light requirements and can not develop well under the broadleaf canopy or in the shade of maturing white spruce. As a result, the white spruces will come to dominate such sites and continue growing over the next 150 to more than 200 years. During this time, white spruces reduce the amount of nitrogen in the soil. Nitrogen is essential to the growth of plants. The roots of spruce trees draw nutrients out of the soil, a perfectly normal process of plant growth in all environments. Such nutrients are then incorporated in the trunks, bark, branches, and the needles of these trees. When the needles are shed, they have a tendency to decay slowly (the thick waxy coating of the needles is at least in part responsible for this). Because of the slow rate of decay in a spruce forest, over a given period of time, fewer nutrients are returned to the soil than are removed from the soil by the root systems of the spruces. Gradual depletion of nitrogen in the soil results, slowing the growth of the spruces and negatively affecting the health of the trees. Weakened spruce become vulnerable to attacks by various insects and diseases. It will take from 125 to 200-plus years for the spruces to die back for lack of appropriate nutrients and because of insect and disease attacks. When the spruces do die, openings occur in the forest canopy, permitting light to reach the forest floor. Sunlight on the forest floor will now reach the layer of mostly undecayed organic matter that was shed by the spruces. Such warming results in an increase in decomposition rates, which, in turn, results in increased release of nitrogen in forms that are usable by any seedlings now trying to develop. The spruces' dying and opening the canopy can also dry out the ground layer of organic matter (forest litter) and mosses. This makes the site prone to the potential of fire and the subsequent invasion of the site by the light seeds of aspen and birch. Even if no fire burns the dried organic materials, breaks in the otherwise closed canopy of spruces permits the light-weight seeds of the sun-loving aspen or birch to germinate and begin the next successional stage. Low nitrogen content of soils can be overcome by both birch and aspen because they have nitrogen-fixing bacteria in small nodules on their roots. Once established on a site, both birch and aspen will continue their relatively short life cycles. These deciduous trees annually shed their nitrogen-rich leaves, the decay of which will again enrich the soil with nitrogen. Thus, these sites once again become suitable for the growth of white spruce, which will invade to repeat the cycle.

Availability of nitrogen in the soil is of supreme importance to the growth of conifers in the Boreal Forest Biome. Low levels of nitrogen or even the absence of this necessary nutrient limits or prohibits the growth and development of conifers. Conifers are not the only plants with a high nitrogen requirement that withdraw as much nitrogen as possible for their growth. Other plants, such as mosses and lichens, also have been proven to be significant contributors to soil nitrogen depletion and thus have been a part of this cycle in which nitrogen depletion results in the dying-off of needleleaf trees. On drier sites, lichens typically form a substantial groundcover under an open or even semi-open canopy of pine or spruce. A layer of light-colored lichens reflects sunlight and thus helps keep the soil temperature lower. This, in turn, causes a decrease in the decomposition rate of organic materials that have accumulated on the soil layer, reducing the nutrients available for plant growth to a level below what is necessary to maintain tree growth.

Mosses are more moisture tolerant than lichens and develop well on forest floors where the forests are closed or where the conditions are simply moister. Mosses do not have roots and a true vascular system. They absorb precipitation through their foliage and, in the process, filter out and trap any nutrients that may be dissolved in that water. As a result, the moisture that does eventually reach the ground through this filter of mosses may not have any nutrients left. The presence of a moss groundcover may speed up the depletion of nutrients in the soil, since the roots of other plants are still removing nutrients. Mosses have other impacts on the soil chemistry. They can absorb and hold so much moisture (up to 4,000 times their weight) that waterlogged areas are formed. Since this water does not contact the atmosphere, anaerobic conditions can develop at the ground surface, prohibiting the decomposition of organic materials that might otherwise recharge the soil with plant nutrients.

Mosses also form what amounts to an insulation blanket over the soil. Soil temperatures are subsequently decreased, which lowers the rate of decomposition of organic materials still further. The nutrient cycle between decaying vegetation and roots of living plants using the products of such decay is thus slowed down or even halted. The total depth of the organic layer on the forest floor becomes thicker as the decomposition of dead needles and dead mosses fails and a litter layer 8–12 in (20–30 cm) thick accumulates—for example, as under stands of black spruce in the interior of Alaska. A thick layer of organic materials on the forest floor retards the regeneration of spruces by preventing the roots of spruce seedlings from reaching the soil layers beneath. Over time, then, the spruce forest dies without regeneration. Its replacement is typically a treeless community of mosses, low shrubs, and herbs. This process is called paludification. It may be the initial stage for subsequent encroachment by sphagnum mosses. Sphagnum mosses by themselves perpetuate a cool, wet, acidic environment. Once established, they may result in the formation of a bog and continue as such with its few shrubs, sedges, and grasses for long periods of time.

Mosses can be significant destroyers of spruce forests in other ways as well. Scientists have found that mosses and the highly acidic soil environments that they promote may indeed kill the fungi that have a symbiotic relationship with tree roots in the extraction of nutrients from the soil. Mosses have been found to create soil water sufficiently acidic to release aluminum from the mineral portions of the soil layers. Aluminum that has been freed from its mineral bond in soil is toxic to most plants and will contribute to trees being killed in the stands so affected. The insulating affect of a thick mat of mosses can also result in a rise of permafrost. When permafrost rises close to the surface, it destroys all but the most shallow-rooted trees.

Animal Life

Throughout the boreal regions of North America and northern Eurasia, animal life is highly dependent on the limited types of food available to them. The needles of coniferous trees are not palatable to most herbivores (except as freshly growing buds when most other edible materials are under deep snow cover), and the nutrient value of the needles is relatively low. Herbivores typically depend on the plants of the forest floor and the broadleaf trees and shrubs growing in and on the edges of bogs, or on sites invaded by deciduous trees after burns or other disturbances. The seeds of conifers are a food source for some animals, and spruce and pine cones can be opened by some mammals and birds to extract the seeds. Cone production, however, varies from year to year and sometimes even within a given region. As a result, animals cannot depend on the availability of seeds in conifer cones and may be forced to migrate south to survive winters when there is a seed shortage in the boreal forest.

A number of mammals and birds remain in the boreal forests during high seed availability, and many of these are panboreal; that is, they are found on both the North American and Eurasian continents in subarctic latitudes. They move south of the boreal forests during times of seed shortages. Some of the better-known animals will be discussed here; the identification of those species limited to one continent or the other will be found in the regional descriptions later in this chapter. The largest and signature herbivore of the boreal forest is the moose, commonly referred to as elk in Europe (see Plate I). This member of the deer family browses on trees and shrubs as well as wetland plants. Moose seem to have a preference for the foliage of poplars and willow trees that commonly grow along the shores of ponds and streams and on the edges of wetlands. The wapiti or elk (see Plate II), which is commonly referred to as red deer in northern Europe, is another large member of the deer family. This herbivore's diet consists mostly of grasses; it is typically seen grazing in open sites of the boreal forest regions where there is sufficient sunlight to encourage the growth of grasses in the ground layer. During the winter months, elk will paw through the snow to the grass layer where possible; if the snow cover gets too deep, they will browse the twigs and foliage of shrubs and trees that extend above the snow. The boreal forests are the winter home of yet another member of the deer family, the barren ground caribou or wild reindeer, as the species is called in the northern sections of Eurasia. When the worst of winter is over, huge herds of reindeer begin their annual springtime migration north into the Arctic tundra regions in both continents. This large migration is one of nature's most photographed, filmed, and recorded spectacles (see Plate III). The annual calving occurs in the tundra, where these herds will feed until late fall, when they will return to the boreal forests for the winter. The woodland caribou is a subspecies that does not congregate in large herds for annual migrations, but rather remains in the woodlands as a year-round resident. All caribou feed on lichens, among which the so-called reindeer mosses seem to be the favorites. During the summer months, caribou feed on the leaves of the broadleaved plants of the tundra. During the winter months, they browse on what lichens and palatable twigs and buds of deciduous trees they can find by digging through the snow. Reindeer, like many other boreal animals, depend largely on energy that they have accumulated and stored as body fat during the summer months. Another large mammal of the northern fringes of the boreal forest regions is the musk ox (see Plate IV).

Among the large carnivores of the boreal forests are the brown bear (also known as grizzly bear), the wolf, and the lynx. The large brown bear is an omnivore and feeds on roots, bulbs, berries, and even grasses and sedges, as well as on fish and meat. Bears have powerful claws for digging roots, but also for excavating the tunnels of voles, mice, and the burrows of hares. In addition, the brown bear is an excellent fisher. Grizzlies have been filmed on numerous occasions catching fish during the times of the year when water is ice-free. Brown bears will take the very young and the infirm from among larger mammals. Other carnivores that are common in the boreal forests of both continents and prey on smaller mammals include the red fox, as well as several members of the weasel family. Among these, the wolverine is a powerful animal that is a scavenger for much of the year, but it is also a skillful hunter. Other weasels include the ermine and the least weasel, which is the world's smallest carnivore. All of these carnivores are furbearers. They and others like them have been long-standing targets of trappers and hunters throughout these regions and have had an important role in the economies of the people inhabiting these forests. To a large measure, it was the hunting of these fur-bearing animals that contributed heavily to the commercial exploration of the boreal forest regions.

Seasonal Changes in Fur Color of Boreal Animals

Several of the animals of the boreal biome molt into different colors, depending upon the season. Such changes are to camouflage them to either (1) make them less visible to their prey, or (2) make them less visible to their predators (and, in some cases, it is a matter of both). Ermine (stoat), arctic wolf, arctic fox, and snowshoe hare are some examples. All of them have a more earthen color (browns, grays, reds) during that time of the year when the ground is not covered with snow. With the onset of cold weather patterns in the fall, they molt their summer fur (the darker hair slowly falls out) and develop a thick winter fur that appears white for lack of color pigments. With the onset of warmer weather, this fur is again replaced with darker and thinner summer fur. The thick winter fur is the reason why these animals are so sought after by hunters and fur trappers.

Of significant ecological importance throughout the boreal forest are small mammalian herbivores such as mice, voles, squirrels, and hares. Any change in their populations has a direct effect on the population levels of the carnivores that depend on them. The periodic population increases and subsequent crashes that many species of this biome experience are one of the best-known and documented features of animal life in the boreal forest. The textbook example used to illustrate this cycle for North America is the 10-year cycle of snowshoe hare and lynx. After fire or other disturbances have removed the coniferous forest canopy over large areas and permitted the establishment of shrubs and broadleaf trees and the growth of grasses as groundcover, food for the snowshoe hare becomes plentiful. The hare, feeding primarily on woody browse, grasses, and broadleaf materials, will experience an increase in its population. Such a population buildup triggers an increase in the population of its chief predator, the lynx. As the vegetation changes due to the natural succession, the food supply for snowshoe hares begins to decrease and, as the lynx population peaks, the death rate among snowshoe hares from the dual influence of predator attacks and a depleting food supply increases, causing a rapid reduction in the numbers of snowshoe hares. The lynx then is faced with a decrease in its primary prey species. To compensate, lynx focus on hunting other species, such as grouse or voles, thereby seriously depleting these animals' populations. Eventually the lynx will face a serious food shortage and its numbers, too, will rapidly decline. This crash in the predator population allows the snowshoe hare to increase again in numbers. When hare are again plentiful, the lynx will shift its hunting back to hares, allowing the voles and grouse populations to recover. The expansion of snowshoe hare populations triggers a rapid growth in the lynx population, and the cycle repeats itself once again.

Several resident birds of the boreal forests are common on both continents, including the Common Raven, grouse such as the Willow Ptarmigan, various woodpeckers, and several owls, including the Great Gray Owl. Songbirds are relatively abundant during the summer months, when most of them arrive from lower latitudes to breed in the boreal forest. Few birds are year-round residents, but they include some chickadees and tits. Redpoll and the Red and White-winged Crossbills depend heavily on conifer seeds as their food supply (see Plate V). As a result, they are wanderers of the boreal forests, always in search of areas with a good cone production. Their residency in the boreal forest depends on the availability of conifer seeds. In years with a sufficient supply, they remain all year. However, in response to the occasionally poor seed years, they undertake erratic mass migrations, known as irruptions, into more southerly regions. When such irruptions occur, crossbills can be found as far south as Virginia and Arizona.

Boreal forests have significant populations of flying insects during the warmer season. The myriad mosquitoes and blackflies are especially aggravating to people and other animals. However, insects constitute the main food item for most of the songbirds that migrate to these northern forests to breed each year. Some insects are considered pests and even an economic threat: foresters have a strong dislike for some of them since they are capable of defoliating and killing trees in large tracts. Throughout the boreal forests of North America, the main culprit doing significant damage to conifers is the spruce budworm, while similarly extensive damage is done to the Eurasian boreal forest by the big coniferous long-horned beetle and the Siberian silkworm. On both continents, these bothersome insect pests go through cycles of rapid population expansions and subsequent decline, similar to the snowshoe hare–lynx phenomenon mentioned above. The rapid growth in the number of these insects seems to parallel the vegetation cycle: they typically attack only forest stands of weakened or dying trees and are rarely effective in defoliating healthy stands.

MAJOR REGIONAL EXPRESSIONS OF THE BOREAL FOREST BIOME

North America

Beginning roughly along a line from the western edges of the Alaska Range to the southern slopes of the Brooks Range and then heading southeastward and eastward across North America, curving around the Hudson Bay and ending on the shores of the Atlantic from Labrador to Maine lies the main area of boreal forest across North America, a nearly unbroken belt of mostly coniferous trees across the entire continent (see Figure 2.4). This belt is at times also referred to as the "interior forest," since most of it is located in the interior of the continent, south of the arctic treeline and away from the coastal zones of the north that are covered by tundra.

In addition to the continuous belt of the boreal forest, there are southward extensions of this forest biome. The most significant variant of the boreal forest is in the Pacific Northwest. Strong maritime influences from the west moderate extreme summer and winter temperatures, and the longitudinal mountain ranges that parallel the Pacific Coast have an orographic effect, resulting in normal annual precipitation levels of 50–70 in (125–175 cm). The montane conifer forests of the Coast Ranges, the Sierra Nevada–Cascade Range, and the Rocky Mountains are considered natural extensions of the belt of boreal forests to the north. Another montane conifer forest extension is in the east and reaches far southward along the spines of the ridges of the Appalachians to the southern portions of the Smoky Mountains, extending discontinuously along the ridges of the Blue Ridge to northern Georgia.

The Primary Belt of Boreal Forests across North America

Throughout the primary belt of the North American Boreal Forest Biome there are nine dominant trees. Six of these are conifers and three are broadleaf trees. The appearance and the structure of this forest belt seem to be rather uniform throughout its entirety. However, there are significant geographic variations in the dominant tree species. In a given genus, species have a tendency to replace themselves from west to east.

In the interior of Alaska, in the northwestern part of the wide boreal forest belt, white spruce and paper birch forests typify the Boreal Forest Biome. In this state alone, they cover nearly 99 million ac (more than 40 million ha). Significant temperature fluctuations characterize the climatic conditions here, with mean annual temperatures of 20° to 30° F (−7° to −1° C), winter temperatures that fall below −40° F (−40° C), and coldest monthly average temperatures of −10° to −20° F (−23° to −29° C). Summer temperatures, on the other hand, may regularly reach above 90° F (30° C), with the warmest month having an average of 60° F (16° C). Precipitation is relatively low: 6–12 in (150–300 mm) with localized variations, and evaporation is low. Permanently frozen ground (permafrost) is somewhat scattered in the central portions of the boreal forest belt, but it is nearly continuous (except on south-facing slopes) in the northern sections and the transition zone to the tundra biome to the north. Permafrost forms an impermeable barrier wherever it occurs, so much of the ground is saturated, and wet areas are common.

The white spruce forests of the western portions of this boreal forest belt typically have a mixture of white spruce, paper birch, and balsam poplar. The best stands of white spruce are on relatively dry hillsides facing south and so having warmed and well-drained soils that are mostly free of permafrost. Mature stands in such areas normally have a relatively open understory, but may contain shrubs of prickly rose, willows, and the occasional alder. The groundcover is often made up of a carpet mat of thick mosses. White spruce can develop on the better sites into fully mature trees that at an age of 100–200 years will reach diameters of 10–24 in (25–60 cm).

Large sections of these white spruce forests in Alaska and northwestern Canada, too, have been burned in recent history. As a result, many portions of these forests are in various successional stages, providing a glimpse of the natural succession of vegetation in this biome (see Figure 2.12). The particular types of vegetation that first become reestablished on a recently burned-over area in the white spruce forests depend on such factors as climate, soils, topography, drainage, previous vegetation, and availability of seed sources. In much of the white spruce forest, fires are typically followed by the establishment of a shrub-willow stage consisting of narrow-leaf Labrador tea, Labrador tea, prickly rose, and willows, all of which have light seeds that are dispersed easily by wind. In upland areas on well-drained south-facing slopes, following a burn and the common willow stage, the

Figure 2.12 Aspen-spruce cycle: young white spruce under aspen. *(© Susan L. Woodward, used by permission.)*

fast-growing quaking aspen often becomes established and matures after 60 to 80 years. Then it is replaced by white spruce (except in areas that are uncommonly dry, where the quaking aspen may persist for much longer periods of time). On reasonably well-drained alluvial terraces, aspen is replaced at times by black spruce.

On those slopes that face east or west, and occasionally even on north-facing and flat areas, the paper birch is the first invading true tree species following a burn, after the usual initial invasion by shrubby willows, Labrador tea, and mountain cranberry. Birches at times become established as nearly pure stands, although white or black spruce may be mixed in. These birches can reach heights of up to 80 ft (24 m) with diameters of up to 18 in (46 cm) at maturity. Balsam poplar is another important tree in the successional stages of the northwestern sections of the boreal forests of North America. It typically invades sites on floodplains that are recently disturbed by fire or other factors and becomes established on reasonably well-drained soils such as sandbars where it grows rapidly to reach heights of up to 100 ft (30 m) and diameters of up to 24 in (60 cm) before white spruce invades and replaces it. In those areas on the floodplains where poplars are the initial dominant tree species, black cottonwood and white spruce typically mix into the association, while American green alder, Sitka alder, willows, prickly rose, and high bush cranberry are common in the shrub layer.

On the northern edges of this belt, on north-facing slopes, and in those lowlands that—because of their geomorphologic history or the presence of permafrost—have poor drainage of their soils, forest succession often leads to the establishment of black spruce stands. The environmental conditions are typically harsh, and black spruce grows slowly. They rarely reach diameters in excess of 8 in (20 cm). Quite often black spruces that measure only 2 in (5 cm) in diameter are more than 100 years old. After disturbance from fire, the black spruce typically becomes established rather vigorously and abundantly: the heat of the fire opens its nearly heat-proof cones and the seeds are spread across the burned-over site. Open black spruce forests on the northern edges of the boreal forests often have as a ground layer a rather thick mat of sphagnum mosses interspersed with sedges and grasses, as well as shrubs such as red-fruit bearberry, Labrador tea, prickly rose, willows, bog blueberry, and mountain cranberry. In the poorly drained bottomlands, rather slow-growing tamarack is interspersed with black spruce, while on the dryer sites, paper birch and white spruce may be found mixed in with the black spruce.

Within the boreal forests are areas that do not support the growth of trees because of their particular environmental conditions. This occurs when the drainage in depressions and flat areas, such as old river terraces and outwash plains, and occasionally on gentle sloping ground that faces north, is so poor that trees simply cannot develop on this water-saturated wet ground. These situations are particularly easily developed when the soil materials contain a high proportion of clay minerals that have a tendency to block drainage. Under these conditions, bogs typically develop. They have varying proportions of grasses, sedges, and particularly sphagnum mosses. One of the curiosities of such bogs is the appearance of an irregular surface consisting of string-like small ridges that are made up mostly of sphagnum moss. For the most part, such bogs do not support the growth of shrubs or woody plants. However, in those sections that are somewhat drier and are basically formed by peat ridges, willow, ericaceous shrubs, and dwarf birches do occasionally grow (see Figure 2.13).

Animal Life

The wildlife of the boreal forests of North America includes a number of animals that are endemic to North America, and many of them are well known. Some were already mentioned above. Woodland caribou and barren ground caribou, moose, and elk are among the large herbivores. Smaller mammals include the herbivores of the tree tops, the red squirrel, and the northern flying squirrel. At ground level, the snowshoe or varying hare, as well as mice and voles search for and feed on seeds, twigs, mushrooms, and fungi on the forest floor. Within the natural food chain, these small animals support a host of carnivores that are capable of moving rather well through the tree canopy: pine marten, fisher, and the long-tailed weasel hunt mostly from the branches and the canopy. Other carnivores, such as lynx, foxes, wolves, and coyotes, hunt expertly on the ground for snowshoe hare and small ground-dwelling rodents, as well as for birds. Along the edges of streams and freshwater lakes and ponds, carnivores that prefer both land and the water environments flourish. Mink are common near open freshwater sources where they pursue fish and crayfish, as well as small mammals such as mice and voles. Rarely out of the watery environment that they

Figure 2.13 Late succession on pond in boreal forest. (© *Susan L. Woodward, used by permission.*)

seem to be most comfortable in, playful river otters hunt for invertebrates, fish, and frogs throughout the streams and freshwater ponds in which they live. The elusive wolverine is a rather powerful animal that is an omnivore during the summer season, but a carnivore and a scavenger during the winter months.

Rodents closely linked to the boreal forests and indigenous to North America include the muskrat and the beaver. Both are commonly found burrowing into stream banks as well as feeding in ponds. The beaver is an expert at constructing its own ponds by damming streams. This animal has significant historical importance in the settlement of North America. It was the premier fur-bearing animal sought by early trappers. The fashionable men's hats made of beaver fur in the Europe of the seventeenth, eighteenth, and early nineteenth centuries brought English and French trappers to the New World and drew them, as well as other explorers, far into the interior of North America in pursuit of this and other fur-bearing animals (see Figure 2.14). The porcupine is yet another rodent unique to North America and a resident of the boreal forest. The porcupine is not one of the favorites among the managers of forest lands: it feeds largely on tree bark during the winter months and causes tremendous damage in commercial forest regions.

A number of both well-known and less-familiar birds are found in the Boreal Forest Biome. The closed-canopy areas of the northern conifer forests as well as the muskegs are home to the Spruce Grouse. The Sharp-tailed Grouse, on the other hand, prefers the open sites of the treeless "islands" in the forest, areas of recent fires, and the drier sections of peat bogs. Several owls hunt the forests stealthily at night. Among

Figure 2.14 Beaver dam in boreal forest. *(Courtesy of Shutterstock #2442827. © Coia Hubert.)*

them are the Boreal Owl and the little Saw-whet Owl. The Barred Owl is at home in the boreal forests regions east of the Rocky Mountains. Several woodpeckers are resident species, among them the very large, red-crested Pileated Woodpecker that is, at times, a loud denizen of the woods, where it feeds on insects. Other woodpeckers include the Three-toed Woodpecker and the Black-backed Woodpecker. These birds are sufficiently powerful to extract dormant insects and their larvae from beneath thick tree bark during the long winter months. This ability provides them with a year-round food supply within the boreal forests and sets them apart from other birds that depend on flying insects only. Another year-round resident of the boreal forest is the raven. A lot of folklore has been woven around this bird throughout North America. It is an omnivore that feeds on carrion to survive the long winter months. A not too-distant cousin of the raven and another resident of the northern forests is the Gray Jay or Whiskyjack. This bird is quite familiar to those who have visited the boreal forest regions, particularly if camping or hiking. It is a relatively tame bird that is a camp follower and frequently is observed scavenging around camp sites.

Most of the songbirds of the boreal forest are migratory. Only about 10 percent of them are actually year-round residents, and most of these are found in the southern areas of the boreal forests. Among these few resident songbirds are the Boreal Chickadee and the Black-capped Chickadee. During the warmer months of the year, many birds migrate into the boreal forests to breed. These include such insect-eaters as the thrushes whose flute-like call can be heard throughout these woods when they arrive in spring. Among the common thrushes are the Hermit Thrush and the Swainson's Thrush. Wood Warblers are represented by more than a dozen varieties and add significantly to the color palette and the variety of songs one can listen to in these woods. Each has adapted to a particular location for nesting and rearing their young. The Palm Warbler typically nests and searches for its food along the edges of bogs. The Bay-breasted Warbler, on the other hand, has a decided preference for the more open coniferous forest areas. Others, such as the Cape May Warbler, prefer a particular type of tree. After their springtime arrival in the boreal forest, Cape May Warblers breed almost exclusively in stands of black spruce. Among a number of other songbirds coming to the northern forests during the warmer months to breed is the White-throated Sparrow. The high whistle of this bird sounds like "Canada, Canada, Canada" and seems to announce in these northern latitudes that the summer months are surely coming. Another breeding songbird is the Purple Finch, whose song contributes to the medley of notes. As mentioned above, birds such as White-winged Crossbills, Red Crossbills, and Redpolls are full-time residents of the northern forests when the supply of spruce, fir, or pine seeds upon which they depend is plentiful. In those winters when this food source is less plentiful, these birds will migrate into the southern ranges of the boreal forest or even farther south in search of their food sources.

By far the most commonly noticed insects are mosquitoes and black flies. Throughout the warmer months of the year, these insects are a tremendous nuisance in all areas of the boreal forests in North America. Foresters, in particular, consider the spruce budworm the most damaging insect. It bores into mostly spruce and leaves the trees open to subsequent damages.

The Evergreen Forests of the Pacific Northwest

The extensive evergreen forests of the Pacific Northwest range from Alaska through western Canada and the northwestern states of the United States into Central California. They are one of the largest, nearly homogeneous forest regions in North America. Extending southward from the northern boreal forest belt these forests have distinctive north-south sections that exhibit latitudinal as well as an altitudinal zonation (see Figure 2.15). While they are all tied to the belt of boreal forests in the north, the individual sections are distinctively different from each other, mostly because of the localized environmental conditions and thus the dominant trees.

The Western Coast Ranges: Sitka Spruce Zone

The Coast Ranges force eastward-bound moist airmasses to lift up and release significant amounts of rainfall on the western slopes of these mountains. Annual precipitation ranges from near 134 in (3,400 mm) near Quinault on the western slopes

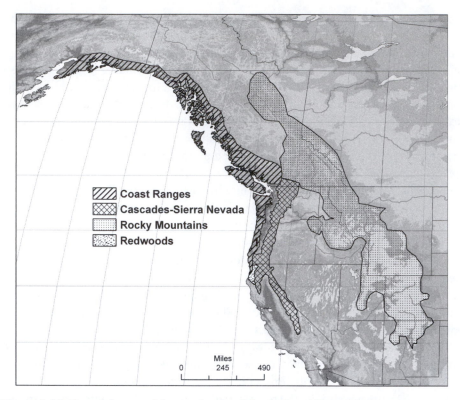

Figure 2.15 Boreal forests of the northwest. *(Map by Bernd Kuennecke.)*

of the Olympic Peninsula to near 70 in (1,800 mm) along the southern Oregon Coast. This section of boreal forest from the coast of Alaska to northern California has been termed a "coastal temperate rain forest."

The eastern slopes of the Coast Ranges are in a slight rainshadow and receive somewhat less precipitation and thus have a slightly different plant association— different enough that some vegetation scientists differentiate between the forests of the western and eastern slopes of the Coast Ranges. The western slopes are the Sitka Spruce Zone, while the eastern slopes are commonly referred to as the Western Hemlock Zone, a zone that ranges across the Puget Sound lowlands and extends up the western slopes of the Cascade Range. Only in the Willamette Valley is the boreal forest interrupted by grasslands and occasional islands of Douglas fir.

Sitka spruce (see Figure 2.16) is regarded as the dominant and characteristic species of a narrow band of forests that reaches from Kodiak Island in Alaska to northern California, where it grades into the forest of the coastal redwoods (see below). This zone is typically only a few to a dozen miles or so in width, except where it can extend inland along larger river valleys or plains. Only on the western and northwestern parts of the Olympic Peninsula, on an extensive coastal plain, is it much broader. Sitka spruce typically occurs at elevations below 465 ft (150 m) and only where mountains rise sharply from the edge of the ocean does it grow higher (up to 1,850 ft

Figure 2.16 Sitka spruce. *(Courtesy of Shutterstock #61914430. © Steffen Foerster photography.)*

or 600 m). At its northern and westernmost limits, on Kodiak and Afognak islands, Alaska, Sitka spruce is the only conifer in this zone. Western hemlock appears in increasing proportions in warmer regions south and east of Cook Inlet. East of Prince William Sound, black cottonwood is a common hardwood along the glacier-fed out-wash rivers on alluvial terraces west of Haines, Alaska. Red alder becomes important along the streams in the central and southern coastal areas of Alaska.

Along the coasts of southern Alaska and British Columbia, forests are composed chiefly of western hemlock and Sitka spruce. Mixed in are mountain hemlock, Alaska cedar, and western red cedar. At low elevations are some fairly large nearly treeless expanses of open muskegs, covered with low shrubs, sedges, mosses, and grasses. An occasional shrubby lodgepole pine, often referred to locally as shore pine, is the only tree. The shrub layer contains Sitka alder, salal, rusty menziesia, devilsclub, currants, western thimbleberry, salmonberry, several willows such as Barclay willow, Scouler willow, and Sitka willow, various blueberries and huckleberries, and high bush cranberry.

In the somewhat warmer environs of the coastal mountains of British Columbia, Douglas fir is increasingly part of the forest mix. This species, characteristic of many of the western extensions of the Boreal Forest Biome, is not found north of the central coast of British Columbia and is completely absent in Alaska. The forests from the central coast of British Columbia to northern California are among the most productive in the world. The dense and tall tree cover is made up most commonly of the three species Sitka spruce, western hemlock, and western red cedar. In addition, Douglas fir, grand fir, and Pacific silver fir are frequent members of the plant association. Lodgepole pine is a common conifer on some of the drier, windblown sites. Red alder, a small deciduous broadleaf tree, grows abundantly on recently disturbed sites (logging activities, wildfires, road construction, and so on).

Toward southwestern Oregon and northwestern California, the transition to the California redwood forest begins. The species of the Sitka Spruce Zone mix with California redwoods, which become dominant farther to the south (see below). Within this transition area, California laurel and Port Orford cedar occur. Mature forests typically have understories that are heavily influenced by the frequent rainfall and commonly high humidity. Dense growths of shrubs, herbs, ferns, mosses, and lichens on the more favorable sites lend a lush appearance. Among the shrubs is quite often red huckleberry, while common plants of the ground layer include swordfern, Oregon oxalis, false lily-of-the-valley, western springbeauty, three-leaved coolwort, evergreen violet, wood violet, Smith's fairybells, and rustyleaf. On those sites where conditions are less favorable, such as on steep slopes with thin soil cover or slopes exposed to ocean winds or on old sand dunes, a dense thicket of shrubs such as Pacific rhododendron, salal, and evergreen huckleberry often develops.

One of the most noticeable aspects throughout the moist regions of the Sitka Spruce Zone of the Coastal Ranges is the relative abundance of mosses and lichens. They both form a continuous groundcover and grow as epiphytes. Red alder is a common host tree for the numerous epiphytes. High rainfall and humidity throughout this "rainforest" let mosses develop in great profusion. In some areas they blanket the ground, extending over fallen trees, as well as up tree trunks and onto the lower branches of living trees (see Figure 2.17).

As in other areas of the Boreal Forest Biome, fire seems to play an important role in natural succession within the Sitka Spruce Zone of the western Coast Ranges. Timber harvesting and other disturbance factors also play an important modern role. Red alder grows rather fast on disturbed sites and reproduces abundantly. It also has some important soil-improving properties through its symbiotic relationship with nitrogen-fixing bacteria. Red alder often grows in association with salmonberry and together they form dense undergrowth that makes it difficult for other plants to become established. Although short-lived, red alder can occupy sites for long periods of time and even semipermanently. Sitka spruce, western hemlock, and western red cedar are often slow to invade red alder stands. Such encroachment commonly occurs on downed logs on which seedlings may become established. Once established, however, the conifers easily outlive the red alder. In the central and southern portion of this zone, the forest of Sitka spruce, western hemlock, western red cedar, and Douglas fir is not the final successional community and is eventually replaced by a single-species stand of the longer-lived and more shade-tolerant western hemlock.

Figure 2.17 Rainforest on the Olympic Peninsula. *(© L. Sue Perry, by permission.)*

Animal life in the Pacific Northwest is closely related to that of the boreal forest belt, although there are some significant differences. Mammals are generally the same or related to those of the continental boreal forest. The caribou of the north is missing in this region; however, another member of the deer family (the black-tailed mule deer) is encountered from southern British Columbia into California. Among the predators, mountain lion and bobcat represent the cat family in this region, replacing the lynx of the north. From Washington southward, spotted and striped skunks are common. The red squirrel is found in the northern parts of this region, from northern Canada to northern Washington. Farther to the south they are replaced by the commonly observed chickaree. Townsend's chipmunk is a resident from California to southern British Columbia. Throughout the coastal mountains of Oregon and Washington the so-called mountain beaver or sewellel is at home. This is the single member of a family of rodents that is endemic to North America.

Sewellel

Sewellel (*Aplodontia rufa*) is a short-limbed, relatively primitive rodent and is the single member of an endemic family. The animal is brownish-red colored on top and grayish beneath. It inhabits moist woodlands in proximity to streams throughout the Pacific Northwest, from northern California to British Columbia. Sometimes called a "mountain beaver," it is not closely related to the beaver. It does resemble a tailless muskrat or marmot. It lives on bark, leaves, and twigs and appears to be rather adapt and gregarious in the construction of complex communal burrows that are typically dug into the banks of streams.

Among birds of the Pacific Northwest the Spotted Owl is perhaps the most famous one. Over the past several decades, this bird has been at the center of controversy. It is a species that is encountered almost exclusively in old-growth forest stands. The bird's population has been declining. As a result, it has become a rallying point for those who advocate the preservation of old-growth forest and its the biodiversity. Other birds of this region typically have a wider distribution. Among them are Varied Thrush and the Chestnut-backed Chickadee. The typically tame Stellar's Jay is found throughout the conifer forests of the western mountains; it is, just like Whiskeyjack cousins to the north, a camp follower and a common thief at picnic tables.

The Eastern Coast Ranges and Western Cascade–Sierra Nevada Ranges

Western Hemlock Zone

From British Columbia to northern California and encompassing most of the Coast Ranges—especially the eastern slopes—clear across the Puget Sound Trough to the western slopes of the Cascade-Sierra Nevada ranges is a subregion of the boreal forest characterized by dense coniferous forests with western hemlock as the indicator species (see Figure 2.15). The wet, maritime climate with a high total annual rainfall amount has an uneven seasonal distribution of precipitation: summers receive only about 6–9 percent of the year's precipitation. The moisture is brought into this region by the prevailing northwesterly weather systems coming in from the Pacific Ocean. While some of the precipitation is intercepted on the western slopes of the Coastal Ranges, the orographic effect of the much higher Cascade and northern Sierra Nevada ranges releases much of the remaining moisture and helps foster a cool moist environment ideal for forest growth.

In the northern sections of the Western Hemlock Zone, annual precipitation reaches slightly over 118 in (3,000 mm), and it averages 59–79 in (1,500–2,000 mm) for most of the region. The Cascade and northern Sierra Nevada ranges receive their winter precipitation as snowfall. The slow melt-off during spring and summer provides the numerous streams of the region with significant permanent flows. In the southern part of this region, conditions are somewhat drier, because the Klamath Mountains of southern Oregon and northern California intercept the airmasses coming in from the west. Actually, a great deal of climatic variation, associated with differences in latitude, elevation, and location in relationship to the mountains typifies this region.

Most of this region has developed on volcanic materials—often in form of thick volcanic ash layers and pumice—ejected through a number of vents (some volcanoes are still active, such as Mount Baker, Mount Hood, Mount Saint Helens, and so on) on top of uplifted older basaltic layers. In concert with the decomposition of organic materials from the dense forests, the moist climate has weathered these materials rather rapidly and created relatively nutrient-rich soils. Decomposition is much more rapid than in the boreal forest belt to the north, where cold temperatures typically retard rates. Although soils are derived from a variety of parent rocks, these soils have some common features. Most soil profiles are at least moderately deep. The acidity is medium. Surface horizons typically are well aggregated and porous, so that standing water is encountered only in low-lying floodplains where meandering streams have left poorly drained depressions. The organic matter content of these soils ranges from moderate to thick. Litter accumulations on the forest floor vary from generally less than 2.75 in (7 cm) to 6 in (15 cm) at higher elevations.

The Western Hemlock Subregion of the boreal forest has a multitude of recognized plant associations that are, to a large measure, tied to the decreasing amount of precipitation received from north to south. The most important representative tree is, of course, western hemlock (although often it is not the most common one), followed by Douglas fir and western red cedar. Grand fir, Sitka spruce, and western white pine occur rather sporadically. In the southern sections, where the precipitation levels are decreased, incense cedar, sugar pine, and ponderosa pine commonly accompany western hemlock on cooler, moister sites. Pacific silver fir grows at higher elevations in the Cascades, from southern British Columbia to northern Oregon. Port Orford cedar is an important species on the eastern slopes of the southern Coast Ranges. Western yew is distributed throughout this subregion, but it is regarded by foresters as an economically unimportant species. Broadleaf deciduous trees are uncommon. They grow in open areas where the coniferous trees have recently been removed by fire or logging, and on wet floodplain areas, where they may form long-lasting subclimax communities. Among the common broadleafed trees are red alder and bigleaf maple; others, such as golden chinkapin occur but are less common.

The initial plant communities in the successional development after an extensive burn or other radical disturbance consist mostly of residual herbaceous plants, such as swordfern, starflower, and whipple vine, as well as invaders, such as woodland groundsel, autumn willow weed, and fireweed. They slowly yield to brush over a period of several years. Shrubs may include Pacific rhododendron, salal, Oregon grape, trailing blackberry, and others. Throughout much of the region, Douglas fir is a seral tree species that becomes established early due to high seed dispersal of its light seeds from neighboring stands. At first Douglas fir grows rapidly in an "even age stand" with a dense canopy that quickly suppresses herbs and shrubs, unless the soils are either very dry or very moist. Douglas fir remains until some canopy openings occur as a result of the natural death of individual trees. Then the true climax species, the highly shade-tolerant western hemlock, appears in the succession and eventually dominates the stand. In wet areas, the successional communities are dominated by red alder, which may continue to dominate through repeated regeneration. Western hemlock and western red cedar are well represented in the early succession on moist but well-drained sites on the eastern slopes of the Coast Ranges. On dry sites, western hemlock does not grow well and will be conspicuously absent from the climax community. Such stands will continue to be dominated by Douglas fir.

The Central Cascades and Sierra Nevada Conifer Forests

Boreal forests cover the higher elevations of the Cascades and Sierra Nevada ranges (see Figure 2.15), where the mountains intercept moisture coming in with weather systems from the Pacific Ocean. A distinct latitudinal zonation of climate occurs. In the northern sections, from southern British Columbia south to the northern half of the Cascades in Oregon, the climate varies from maritime cold in the north to maritime cool in the south. While rain falls throughout the year, summer months receive only 15 percent of the annual precipitation.

Farther south, in the volcanic Cascade Range and through the granitic Sierra Nevada Range, mediterranean climate patterns with their typical summer drought periods become an increasing influence even in the high mountains. The central and southern sections, along the backbone of the Sierra Nevada, typically receive more than 50 percent of their annual precipitation during the winter months, primarily as snow. At the upper elevations, near the crest line of the northern Cascades and the northern extension of this mountain chain into southern British Columbia, precipitation levels of nearly 78 in (2,000 mm) are common, while the lower slopes at the southern end of the Cascades may receive only 10–14 in (25–35 cm) of annual precipitation. Average temperatures also vary from north to south, as reflected in the elevations of treeline on these mountains. In the north, many peaks extend above treeline, which is at about 6,500 ft (2,000 m). Farther south, with warmer average temperatures, the treeline climbs to about 11,500 ft (3,500 m).

The plant associations mirror the climatic and the altitudinal changes. Species composition varies from north to south and from the lower western slopes to the high alpine backbone of these mountain ranges and immediately east of the crestline, where the orographic effect of the mountain range is reduced. In general, these alpine forests can be classed into two different zones: a subalpine zone and a montane zone (see Figure 2.18). The higher one is the subalpine zone, which reaches

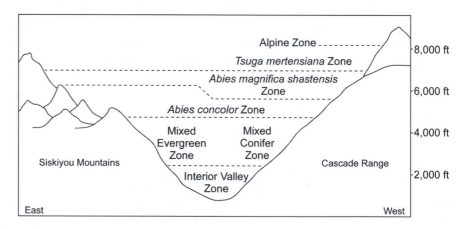

Figure 2.18 Altitudinal zonation from the Siskiyou Mountains to the Cascades in southwest Oregon. *(Illustration by Jeff Dixon.)*

from 4,900 ft (1,500 m) to right below treeline (above 6,600 ft), skirting the crest of the mountain chain. From southern British Columbia through Washington and south into Oregon as far as Crater Lake, the subalpine zone is subdivided into Pacific silver fir and mountain hemlock subzones. The Pacific silver fir is located west of the crest line. The association typically includes Pacific silver, western hemlock, noble fir, Douglas fir, western red cedar, and western white pine. Localized species of importance can include subalpine fir, grand fir, Engelmann spruce, lodgepole pine, and western larch. Mountain hemlock and Alaska cedar grow in the higher elevations within this subzone. The shrub layer is dominated by heaths, such as huckleberry, salal, rhododendron, and dwarf blackberry. Evergreen violet and beargrass are among the plants in the herb layer. Following severe fires, the succession of trees begins typically with either Douglas fir or noble fir, or both. Western hemlock may come in the early stages as well, or may join later. As the stand matures, neither Douglas fir nor noble fir regenerate, because neither tolerate shade. The real climax species of this zone, Pacific silver fir, typically develops under the canopy of the initial stands of Douglas fir, noble fir, and western hemlock. If the disturbed site is small, and if Pacific silver fir is present in neighboring stands, then this conifer also may be represented during the earliest stages of secondary succession.

The Mountain Hemlock Subzone covers the upper reaches of the subalpine zone along the western edge of the slopes. While named after mountain hemlock, this subzone typically does not have a single dominant species. The characteristic tree associations include mountain hemlock, grand fir, lodgepole pine, Pacific silver fir, Alaska cedar, Douglas fir, and western white pine. Proportions of these trees change according to the local climatic characteristics. The volume of winter snow seems to have a significant influence on the vegetation composition. The variety of trees in this subzone is mirrored in the number of species in the shrub layer. Dominant are members of heath, rose, and sunflower families. The early successional stage is often characterized by the widespread dwarf huckleberry and beargrass, which can form a complete surface cover and actually persist for long periods of time. Both are resistant to the frequent wildfires that occur in this subzone and often kill tree seedlings and young saplings. When trees do become established, the first to colonize moist soils are mountain hemlock and Pacific silver fir, while lodgepole pine and subalpine fir are the first trees to invade drier sites. During the subsequent development of forest stands, the number of species increases until the abovementioned variety persists as a climax community. South along the backbone of the Sierra Nevada in California mountain hemlock and lodgepole pine continue to dominate the subalpine zone. Whitebark pine dominates on dryer sites, while Shasta red fir grows where more moisture is available.

Throughout the higher sections of the montane zone of this region, Douglas fir and white fir dominate most stands. Incense cedar and sugar pine are relatively common on moister and drier sites, respectively. The lower montane conifer forests of the Sierra Nevada are often dominated by Jeffrey pine; at even lower elevations and lower precipitation levels, ponderosa pine characterizes these forests.

Sequoia, Redwood, and Western Red Cedar: Forest Giants

The terms "sequoia," "redwood," and "red cedar" seem to be somewhat confusing in the popular language of the trees of the western United States. The case of western red cedar (*Thuja plicata*) is relatively simple: it is a tall conifer that is a member of the cypress family (Cupressaceae) and an important timber tree because of its straight-grained wood. Standing 150–250 ft (50–80 m) tall, it is found in the northwestern United States and Canada, from southern Alaska to northwest California and east to the Rocky Mountains in western Montana and throughout Idaho. (The red cedar of the eastern United States is a misnomer. This tree is actually a juniper.)

Giant sequoia and redwoods are also members of the cypress family. Giant sequoia (*Sequoiadendron giganteum*) is the single species in the genus *Sequoiadendron*. It has scaly needles resembling a cedar or juniper. Sequoias grow only in the western Sierra Nevada of central and south-central California, where it is preserved in the Giant Sequoia National Monument, Sequoia and Kings Canyon national parks, Yosemite National Park, and neighboring forest regions. This tree is a true forest giant with an average height of 165–280 ft (50–85 m).

Coast redwood or California redwood is a member of the genus *Sequoia*. Its needles (see Plate VI) are similar to those of hemlocks or Bald Cypress, an eastern deciduous member of the cypress family. Redwoods (*Sequoia sempervirens*) are common only in the coastal region of northern California (extending into southwestern Oregon). Redwoods National Park and the numerous protected groves of these forest giants, which attain heights of more than 330 ft (100 m), are an attraction for thousands of tourists each year.

In the central and southern Sierra Nevada, localized higher rainfall levels result in pockets of moisture in an environment that is otherwise characterized by a mediterranean climate. High-reaching valleys seem to trap moisture borne by westerly winds from the Pacific. The orographic effect of the high mountain crest to the east causes clouds to mass on the windward side and precipitation to fall in amounts necessary for the development of a number of scattered groves of giant sequoia. Among these groves is the singular specimen that has been named "General Sherman." It has a tremendously large biomass, the very reason it has been called the world's largest living organism. The tree has a height of 275 ft (84 m), and above its roots, it is 103 ft (31 m) in circumference. The total volume of the tree's trunk has been estimated to be about 52,500 ft^3 (1485 m^3). It is estimated to be 2,500–3,000 years old.

The Redwood Forests

The redwood forests of the Pacific Coast region grow in the Klamath Mountains of southwestern Oregon and the Siskiyou Mountains of northern California, as far as south-central Humboldt County, California (see Figure 2.15). These forests receive their name because significant proportions of the forests are made up of coastal redwood. This forest zone at the southern end of the Boreal Forest Biome only reaches inland a little over 15 mi (about 20 km) from the coast. Some question whether this is indeed a separate forest zone or if it is a part of the Sitka Spruce Zone, the Western Hemlock Zone, or even a part of the mixed conifer forests of Sitka spruce, western hemlock, Douglas fir, and grand fir that are found between the Coast Range and the Cascade Range.

Coastal redwood forests develop on mountain slopes. The giant trees can reach ages in excess of 1,000 years. They are probably not the true climax trees (which may be western hemlock and tanoak), but they often seem to remain dominant "forever" due to their longevity. A low replacement rate could perpetuate the stand for centuries or even millennia. On dry sites, the major associations include Douglas fir and tanoak with Pacific rhododendron and evergreen huckleberry in the shrub layer. On more moist sites, redwoods are commonly associated with Douglas fir, California laurel, and bigleaf maple. Evergreen huckleberry is a common shrub, and the ground-cover includes swordfern and Oregon oxalis. The successional stages are poorly known. Mature coastal redwoods are fire resistant, but fire seems to be required for seed propagation and survival of the species. Some evidence suggests that propagation is associated with periodic sheet flooding of the slopes.

The Rocky Mountain Conifer Forests

The conifer forests of the Rocky Mountains are a second southward-tending branch off the main boreal forest belt. These montane forests range from western Alberta across the continuous mountain chains commonly referred to as the Northern Rockies and Central Rockies into Colorado, from where they spread by way of the more separate peaks of the Southern Rockies into Arizona (see Figure 2.15). The coniferous forest associations of the Rockies display a latitudinal zonation, similar to the Cascade–Sierra Nevada ranges to the west. However, three distinctive altitudinal zones exist in the Rockies: alpine, subalpine, and montane zones (see Figure 2.19).

Compared with the mountains to the west, the climates of the Rockies are more continental, particularly at the higher elevations. Temperatures can be much lower

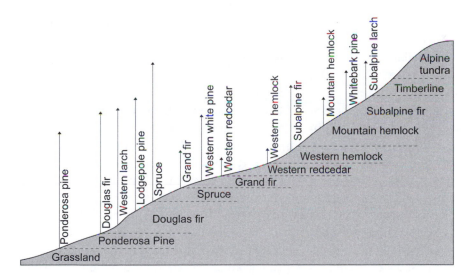

Figure 2.19 Altitudinal zonation of needleleaf trees in the Northern Rocky Mountains of Western Montana. *(Illustration by Jeff Dixon.)*

because of the much greater height of these mountains. Precipitation varies from 15 to 60 in (40 to 150 cm), but can be higher locally because of the orographic effect of these mountains. A large portion of the annual precipitation occurs as heavy snowfall during the long winter months. This has, of course, a severe physical impact on the forest vegetation, favoring those trees that are able to shed the snow-load and helping to retard the growth of those species that suffer extensive limb breakage. Englemann spruce and subalpine fir are dominant trees in the subalpine zone, as they are in the subalpine zones of the Cascade Range, the Olympic Range, and the mountains of southern British Columbia. Bristlecone pine, which is reputedly one of the oldest living organisms, often joins them at treeline. Fir, spruce, and pine usually exhibit a krummholz growthform in the upper subalpine and lower alpine regions of these and other mountains of western North America close to the respective treeline. These stunted trees grow in the windshadow of boulders and often have most of their branches close to the ground, as well as a spindly growth that frequently does not reach more than a few inches above the average depth of the winter snow cover. As in other boreal forests, wildfires occur regularly during the drier summer months. The succession that follows such disturbances is similar to that already described for the Cascade–Sierra Nevada ranges. Throughout the altitudinal zones of the coniferous forests of the Rocky Mountains, when wildfires or avalanches destroy fir and spruce stands (see Plate VII), extensive even-age stands of lodgepole pine and quaking aspen often develop. Where fires are frequent, these stands persist for long periods of time before Engelmann spruce and subalpine fir are reestablished. Downslope from the subalpine zone, Douglas fir is the dominant species in the upper montane zone. The orographic effect of the mountains lessens as elevation decreases, so that the lower montane regions of the Rocky Mountains are drier. Lower total precipitation favors the development of semi-open forest parklands dominated by ponderosa pine, as is the case on the eastern slopes of the Cascade–Sierra Nevada ranges (see Figure 2.20).

The wildlife of the boreal forests of the Rocky Mountains includes most of those encountered in the mountains farther west and north. Large herbivores are bighorn sheep, wapiti, moose, and mule deer. Moose typically is found only in the Northern Rockies of the United States and Canada. Other large mammals with relatively even distribution throughout the region are black bear and mountain lion. The grizzly or brown bear is restricted today to the Northern Rockies of the United States and Canada, where it is protected. As is common in the Boreal Forest Biome, a variety of squirrels take advantage of the availability of seed cones. Among them are the red squirrel, the tassel-eared Abert's squirrel, the golden-mantled ground squirrel, and the least chipmunk. The last two have little fear of people and are commonly seen begging for handouts from visitors throughout the many national parks of this region. Other small animals are not quite so friendly. Beaver, muskrat, snowshoe hare, and porcupine are other relatively common occupants of the Rocky Mountain forests and wetlands. The birds of the Rocky Mountains are basically the same as those encountered in the Cascade–Sierra Nevada ranges.

Figure 2.20 Parks dominated by ponderosa pine in the Rocky Mountains. *(© L. Sue Perry, by permission.)*

The Appalachian Mountain Conifer Forests

The Appalachian Mountains are old mountains that consist largely of sedimentary rock formations that have been folded and faulted. A few sections, such as the Blue Ridge Mountains, are geologically different in that they consist largely of igneous rocks, some of which are the Precambrian granitic basement rock of the continent, but most of which are younger materials, injected into the older sedimentary rock and subsequently uplifted and sometimes faulted and folded.

The Appalachian Mountain conifer forest in eastern North America is another branch of the boreal forest belt. This southern extension follows the eastern mountain chain at higher elevations from eastern Canada to New Jersey, through the western sections of Virginia to West Virginia, and on to the southern Appalachians (see Figure 2.21). The elevation at which this coniferous zone occurs increases from north to south. On the mountains of New Hampshire and Vermont, the boreal zone lies between 2,600 and 4,000 ft (800 and 1200 m). On the high plateaus of New York, it occurs above 3,600 ft (1,100 m). Near the southern margins of the forest, in the Great Smoky Mountains of Tennessee and North Carolina, the conifer forest zone begins above 4,500 ft (1360 m). Precipitation increases from north to south, but the snow amounts are less to the south and the duration of snow cover is relatively short.

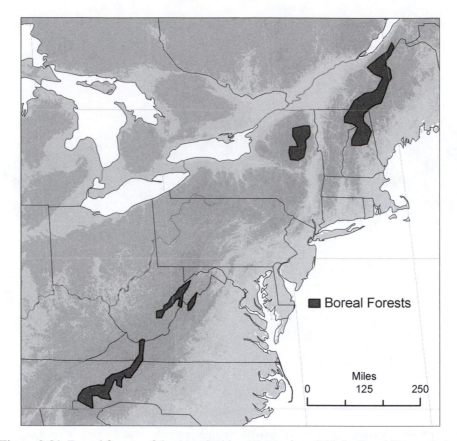

Figure 2.21 Boreal forests of the Appalachian Mountains. *(Map by Bernd Kuennecke.)*

The plant associations in the northern Appalachian Mountains are basically the same as in the eastern portions of the northern belt of boreal forests. Balsam fir and red spruce are common conifers, while paper birch is a frequent broadleaf deciduous tree (see Figure 2.22). The boreal forest along the spine of the Appalachian Mountains is not continuous and, this is especially so in the southern part of the mountain chain, where it is often interrupted by long distances of temperate broadleaf deciduous trees. As a consequence, none of the species of the northern sections have ranges extending the full length of the Appalachians. Both balsam fir and paper birch have their southern limits in the Shenandoah National Park, in the Blue Ridge Mountains of northwestern Virginia.

There is a gap of more than 100 mi (160 km) before another fir forest occurs. Such a geographic separation seems to have been sufficiently great and has persisted long enough for a different species to have evolved and characterize the boreal forests of the Southern Appalachians. On Mount Rogers, Virginia's highest peak (5,597 ft or 1,696 m) and close to the Virginia–North Carolina state line, Fraser fir provides a crest-line forest cover. Fraser fir is the only fir to the south, in

Figure 2.22 Paper birches. (© *Susan L. Woodward, used by permission.*)

the even-higher Great Smoky Mountains. South of the Shenandoah National Park, paper birch is replaced by yellow birch. In contrast to balsam fir, which is found only in the highest elevations and is discontinuous in distribution, red spruce typically grows at lower elevations from Canada to the southern sections of the Appalachian Mountains. It is much more widespread in the Middle and Southern Appalachians than either fir. Throughout the Appalachian Mountains, an understory consisting of heaths such as azaleas, rhododendrons, and huckleberries is rather common beneath a canopy of red spruce. Various mosses, clubmosses, and lichens typically make up the plant associations of the ground layer beneath spruce as well as spruce-fir forests throughout the region. The boreal forests of the Appalachians cover a relatively small area when compared with the total size of the mountainous region, and they are regarded as highly endangered ecosystems.

The wildlife of these eastern mountains differs from that of the western part of the continent. Far fewer large mammals inhabit the Appalachians, in part because the more populated east has experienced much higher hunting pressures. Habitat change, in particular, the changes brought about by agricultural and timber harvesting, is also implicated in the loss of species. The Appalachian Mountains are still home to deer, black bear, mountain lion, moose, and even elk that are being reestablished in some areas. Moose is only encountered in the northern sections, though their range is expanding. The white-tailed deer plays the same role as its counterpart, the mule deer, in the west and is abundant throughout the Appala-chians (some would say too abundant). It has reached population levels that make it a pest on farms and in suburbia. Among the smaller carnivores restricted to coni-fer forests in the northern sections of the Appalachians are pine marten and fisher. Snowshoe hare and the flying squirrel are similarly confined to spruce and spruce-fir forests. The long-tailed weasel and the red squirrel, though characteristic, are ecologically more widespread and therefore have a wider geographic distribution.

The birds, too, are different from those of the western mountains. However, there is considerable overlap among the migratory species that breed in the boreal forest belt. In high elevation conifer forests of the eastern mountains are White-throated Sparrow, Golden-crowned Kinglet, Ruby-crowned Kinglet, Dark-eyed Junco, and Purple Finch, as well as some of the wood warblers and thrushes. Sev-eral of the resident birds that are the same as in the northern boreal forest belt are Black-capped Chickadees, Red-breasted Nuthatches, and Common Ravens.

Eurasia

The Primary Belt of Boreal Forests across Northern Eurasia South of the Zone of Permafrost: The Taiga

Across northern Eurasia from the western coast of the Scandinavian Peninsula to the eastern coast of Siberia, stretches the other large belt of the Boreal Forest Biome (see Figure 2.23). East of Scandinavia, this wide belt of northern needleleaved trees south of the zone of permafrost is often referred to as the taiga (see Figure 2.9).

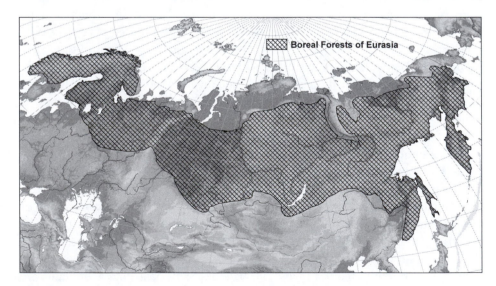

Figure 2.23 Boreal forests of Eurasia. *(Map by Bernd Kuennecke.)*

Throughout this vast region only 14 kinds of trees have been identified dominating the boreal forests. Ten of these 14 are coniferous trees, while four are broadleaf deciduous trees. Although each may be dominant in one or more geographic regions, none ranges the full breadth of the belt. Since each is spatially limited, distribution patterns of these trees help delineate the several subregions within the Eurasian part of the Boreal Forest Biome.

Two subregions occupy western Eurasia: the Fennoscandian subregion, which includes northeastern Norway and Sweden, as well as western Finland; and the European Russia taiga subregion, which includes a Karelian taiga subsection in the north. East of the Ural Mountains in Siberia (the traditional dividing line between Europe and Asia) lie the Western Siberian taiga subregion and the Central and Eastern Siberian subregions. The eastern part of Eurasia's boreal forest is not homogeneous, and forests are denser in the east than in the west. Indeed, it is highly differentiated according to local environmental conditions (climate, surface configuration and composition, and surface and subsurface water levels).

The Fennoscandian Subregion

The Fennoscandian subregion of the Boreal Forest Biome geographically corresponds to the Scandinavian Peninsula plus Finland, and it encompasses the boreal forests of Norway, Sweden, Finland, and the northernmost portions of Russia located on this peninsula (see Figure 2.23). Two species dominate these conifer forests: Norway spruce, which prefers the slightly moister sites, and Scots pine, which has a preference for the somewhat drier sites. However, both trees have broad habitat tolerances. Norway spruce dominates those geographic regions where the annual temperatures as well as the humidity levels are modified and raised by proximity to the sea. It also prefers sites where frequency of wildfires is relatively low. In contrast, Scots pine dominates the more continental sites, which have lower annual temperatures and slightly less moisture, and are more prone to wildfires. Both spruce and pine can reach a significant age. The upper age-range for Norway spruce is more than 400 years, while Scots pines may attain an age of more than 700 years.

Broad habitat tolerances of the dominant trees, significant local climate variations, and mosaics of soil conditions, varying disturbance history, and age structure all contribute to a great variety in the structure as well as composition of these boreal forests. Pine forests often exhibit mature trees of significantly different ages in the same stand. This indicates that the largest trees have survived frequent fires and have contributed to regeneration of the stand. In contrast, old-growth spruce forests for lack of fires have few large-diameter trees and a large number of smaller, stunted trees that have not been "thinned."

In the mountainous sections of this subregion, subalpine forests occur. Birch often dominate these plant associations. Frequently, other trees join birch—for example, mountain ash, dwarf birch, and juniper. Throughout the Fennoscandian subregion few shrubs are associated with the forests compared with other boreal areas. While tree saplings do occur as a shrub layer, shrubs are more commonly associated with successional stages and include willow and juniper. On moister sites, goat willow and rowan may grow as an understory within the Norway spruce-dominated forest stands. Both shrubs are highly palatable to mountain hare and moose and subsequently heavily browsed. As a consequence, they rarely attain actual tree size. Dwarf shrubs such as bilberry and cranberry also grow in moist locations. On drier sites and in the pine forests, the shrub associations commonly consist of heathers. The groundcover of the pine forests is mostly lichens. These lichens provide important winter forage for the domestic reindeer herds that are found in northern Finland and in the neighboring areas of northeastern Norway, Sweden, and northern Russia. A distinctive south-north gradient in the occurrence of herbs and shrubs in the Fennoscandian region is evident. Herbs are more common in the south where their habitat requirements are commonly met. With increasing northern latitude, more and more evergreen dwarf-shrubs make up the ground layer.

The animal communities of the Fennoscandian subregion include moose, reindeer, and mountain hare. Roe deer is also present. Both roe deer and moose have experienced significant changes in their populations over the past few decades. Because of the hunting pressures exerted upon these animals in the 1970s and 1980s, their populations were declining. Decreasing cattle grazing in forests and an increasing food supply as a result of clear-cutting (and the resultant establishment of successional herb and brush species while reforestation was in progress) have since provided more habitat for these herbivores, so deer and moose populations have rebounded in recent decades. Furthermore, a drastic decline of red foxes, a major predator of roe deer, due to diseases has contributed to the rapid increase in red doe populations since the late 1980s. Mountain hare is the preferred prey of the red fox. The relationship of these two animals cycles through the population fluctuations that seem to be characteristic of some boreal forest animals. Other significant changes in the mammalian fauna during the past decades include the accidental or deliberate release of some nonnative species, such as the North American muskrat, the American mink, and the raccoon dog from East Asia. All are negatively affecting the natives of the region, particularly bird populations. Many of the birds of the boreal forest belt of North America are also found here. The diversity and abundance of birds decrease with the increasingly lower temperatures of the northern and eastern sections. Exceptions occur wherever significant numbers of deciduous trees mix into pine or spruce stands. The greater variety of seed sources thus provided may permit larger and more diverse bird populations. Noteworthy are the endemic members of the grouse family in these forests. Among them are the large Capercaillie, Siberian Spruce Grouse, Black Grouse, and Hazel Hen.

The European Russia Subregion

The boreal forest of the European Russia subregion is often referred to as the taiga (see Figure 2.23). Its northern extension, the Karelian taiga, grades into a tundra ecotone, where the dominant Siberian spruce and some birch forms a relatively open forest. To the south and east of the Karelian region, the taiga is a much more heavily forested region dominated by Scots pine. A significant gradation in stand density ranges from thinly forested areas in the north near the boundary with the tundra to heavily forested stands in the central portions. Lower stand densities again characterize the southern edge of the taiga. While Scots pine dominates stands throughout this subregion, spruce and birch are nearly as common and are dominants in many areas. Siberian spruce is common in northern areas, as is Siberian larch. In the northeastern sections of the taiga, Siberian fir and Siberian cedar pine are prevalent and dominate many stand areas. The boreal forests of European Russia are often referred to as a "green moss" community. Typically, relatively few shrubs and herbs occupy the understory, but an ample growth of "feather mosses" cover the ground.

To the east, in the northern areas of the Ural Mountains, is a boreal forest in which Siberian larch, Siberian spruce, and birch are the dominant trees. At the upper elevation limits of the boreal forest in these mountains, near treeline, trees give way to a thick shrub cover consisting largely of Manchurian alder. Siberian larch in these mountains grows in a variety of environmental settings, ranging from lowlands and river bottoms up to treeline.

The Western Siberia Subregion

Boreal forests east of the Ural Mountains form the western Siberian subregion of the biome (see Figure 2.23). A vast geographic region, western Siberia is largely a geologic depression transformed during the Pleistocene Ice Ages into an undulating plain filled with glacial materials and landforms such as ground moraines, till plains, eskers, and drumlins. As a result, much of the region is poorly drained. Numerous bogs, large and small, cover thousands of square miles. As Pleistocene ice sheets retreated, glacial meltwater dammed-up by the ice along the northern edges of this region created many lakes. Today, the climate is basically continental and precipitation normally exceeds evaporation due to low temperatures. Furthermore, abundant surface water is available. As a consequence, flooding is common throughout most of this region in the spring, when snowmelt from the southern portions of the drainage basins is blocked by still-frozen sections near the mouth of the north-flowing rivers. Large bog areas cover approximately one-third of this region at least part of each year. The subregion is, for the most part, relatively heavily forested. The tree associations are dominated primarily by Siberian larch, Sukachev's larch, Siberian cedar pine, and Siberian spruce. One of the characteristics of the boreal forests in this subregion is the continuous mosaic of nearly single-species stands of the dominant Siberian larch or Sukachev's larch that are only sometimes intermixed with Siberian spruce in the north. The higher, better-drained sections of the undulating Siberian plains often have sandy soils. Here lichens form a continuous groundcover.

Central and Eastern Siberia Subregions

The Central and Eastern Siberian subregions of the Eurasian Boreal Forest Biome (see Figure 2.23) have landscapes of largely low hills and well-rounded mountains, the results of sculpture by Pleistocene continental glaciers. As a consequence, these boreal forests are much better drained than those in western Siberia, and environments range from moist floodplains along streams to thin-soiled uplands in the mountains. These regions have a much colder and drier continental climate than western Siberia. Siberian spruce and Siberian cedar pine are coniferous trees that survive such climatic conditions. East of the Yenisei River, which forms the boundary with the western Siberian subregion, spruces begin to thin out and then disappear until they reappear near the continental edge along the Pacific Coast. A region of some 950,000 mi^2 (2.5 million km^2), this vast area is dominated mostly by larch. In the northern sections, relatively open stands of Dahurian larch, resembling the North American larch in some ways, are able to withstand the severely cold temperatures of winter. The Dahurian larch grows on a variety of soil types. In its northernmost locations, larch typically reaches heights only slightly above 3 ft (1 m) and is associated with the occurrence of permafrost. The relatively shallow root systems of larch permit it to grow in the thin active layer that thaws at the top of the ground during the warm season of each year; otherwise only the most dwarfed trees, such as Japanese stone pine, can tolerate such extreme ground conditions. In some areas, even dwarfed larches cannot gain a foothold.

Away from the northern fringes of the subregion in its central area, Siberian larch does become a forest tree of significance. On sites where optimum conditions for its growth exist, this tree can gain heights of 150 ft (45 m) or more and have a diameter of 6 ft (1.8 m). Farther south, with both increasing elevation and drier sites, Siberian pine becomes increasingly important. Siberian pine may appear in pure stands or mixed with spruce, larch, fir, cedar, birch, and aspen, depending on environmental conditions. Toward the southern transitional zone (ecotone) are localized pure stands or mixtures of birch, aspen, and extensive mixed stands of Siberian larch and spruce, as well as associations of pine and larch.

The boreal forests of Central and Eastern Siberia continue eastward into northeastern China and northern Japan. The vegetation of this eastern subregion ranges from an open forest-tundra ecotone of open larch forests in the north to the central and southern areas, where Siberian pine, fir, cedar, and spruce assume increasing importance as dominant trees. The community of plants at the ground layer resembles the composition and structure of the western areas of the boreal forest in North America.

The wildlife of these eastern regions of the Eurasian boreal forest belt is similar to that found in North America. Panboreal animals were mentioned above. The important fur-bearing species are close relatives (congeners) of American weasels and rodents, and like them, contributed significant to the exploration of this remote part of Eurasia. Endemic fur-bearers of the central and eastern Siberian subregions of the boreal forest include the European mink, the European pine marten, the European beaver, and the sable. The sable played a role in central and eastern European political history, since its fine furs were worn exclusively by royalty as a sign of their status. Other animals that resemble some of those of North America, but are only cousins, include the river otter and the flying squirrel.

The resident birds of the eastern sections of the Eurasian boreal forest belt are similar to those of the boreal forest of North America. However, lacking are many of the summer migrants, including the colorful Wood Warblers that live and breed in the northern coniferous forests of Canada and the United States. Tits are found here as in North America, where they are called chickadees. Representatives of this family of birds are the Willow Tit and the Siberian Tit. The range of the latter extends from Eurasia across the Bering Strait into Alaska. Endemic to Eurasia are representatives of the grouse family, including the rather large Capercaillie and Siberian Spruce Grouse, as well as a more normal-size grouse, the Hazel Hen.

Human Impacts and Contemporary Conservation Issues

Significant sections of the circumpolar boreal forest remain to this date largely intact in terms of their ecological diversity and composition. However, if past and present indicators on both continents are taken into consideration (the results of trapping, hunting, logging, and mineral exploration and exploitation), then indeed there have been and still are significant impacts from human activities in these regions. Modifications as a result of human activities will continue to change the Boreal Forest Biome to an ever-increasing degree. Hunting and trapping have been major agents of change on both continents, altering the diversity and population characteristics of predators and their prey, and the role the animals play in seed dispersal and plant successions. The survival of many people living in the boreal forest regions still depends to some degree on their ability to hunt and trap fur-bearing animals of the boreal forests.

Logging has been and will continue to be a major agent of change on both continents. The removal of trees in conjunction with commercial harvesting activities has been ongoing and has passed through several different stages. The initial logging in boreal forests was triggered by the desire to harvest the larger and taller pines, spruces, and tamaracks for saw-timber, to be used for construction within the region and also for export to the more populated parts of both continents. The growth of the large trees to saw-timber size is slow in these northern environments. Logging has changed its orientation increasingly toward harvesting spruce to fulfill the ever-growing demand for pulpwood by the paper and cellulose industries.

Current practices of forestry—for example, in the highly mechanized and efficient Fennoscandian forest industry—have had a tremendous impact on the biodiversity of the boreal forests in Scandinavia. The loss of natural forests and their replacement with managed commercial forest stands has negatively affected both plant and animal populations. In some cases, efforts are under way to moderate the impact of modern forestry so that the natural biodiversity of the region can be reestablished and maintained on a sustainable basis.

Modern logging needs to operate economically, and clear-cutting is the most cost-effective way to harvest the large stands that are made up of only one or two species (see also Plate VIII). Cut-over areas with good soils are quickly replanted with conifers, by-passing the natural vegetation succession that would involve a broadleaf stage in the natural renewal of the forests. Once replanted, such artificial stands are a significant part of the modern emerging landscapes of the boreal forests: Carefully managed, even-age stands contain only those species that have an important role in the forest products industry. Evidence of such changes within the boreal forest belts can be seen in North America from northern New England, Eastern Canada, the Upper Peninsula of Michigan and neighboring Wisconsin, to the northern edge of the Rocky Mountains in Canada, and in central Alaska. Similar observations can be made from Finland to the Urals in western Eurasia and, to a lesser extent, east of the Ural Mountains to the Pacific shores.

Logging of the tall trees from the boreal forests has resulted in the removal of much old-growth forest. This has led to a decrease in the number and types of those plants and animals that were habitat specialists in old-growth forests, and at the same time, has allowed an increase of generalists that prefer disturbed sites. In those cases in which replanting has occurred with only desirable plant species, the plant diversity has been severely reduced. Studies of forestry practices in the northern parts of Finland clearly demonstrate that among the old-growth specialists that are threatened by modern forestry practices are lichens, orchids, Siberian Tit, marten, and Capercaillie. Closer to home in North America, in the Pacific Northwest, the debate surrounding the Spotted Owl and its habitat in old-growth Douglas fir forests has captured significant attention. Yet, the Spotted Owl is but one of a number of species threatened by the logging of old-growth forests in the Pacific Northwest.

Other human impacts are closely related to hunting practices and policies. The Willamette Valley of Oregon, when discovered by members of the Lewis and Clark Expedition, was an open grassland with islands of trees. Contemporary research finds that the removal of the tree cover in that valley was the result of Indian hunting practices that used fire. Throughout Scandinavia and northwestern Russia, wolves, brown bear, wolverine, and lynx have become rare. Their populations are intentionally kept to low levels or are eradicated, as in the case of wolves in much of Scandinavia, by hunting practices and policies. The intentional reduction of predators has, as least in part, allowed populations of moose and roe deer to increase dramatically in the same region. Other factors also come into play. Cattle grazing in the boreal forests of northwestern Eurasia used to be somewhat important. It was carried on as an extensive type of land use throughout the forest regions here. Reduction of the practice of grazing cattle in forests during the past three decades has resulted in an increase of forage for other herbivores. In addition, stands of trees that have been clear-cut offer an increased amount of forage the first few years after cutting. Combined with the hunting practices and policies keeping the population levels of predators very low, clear-cutting seems to favor an expansion in the populations of large herbivores such as moose and roe deer.

Recent drainage of wetlands and bogs in conjunction with agricultural expansion, peat mining, and drainage of peat bogs to help the reestablishment of commercial forest areas all have had a tremendous impact on the ecological characteristics of bogs, particularly so in the southern sections of the boreal forests regions.

The introduction of nonnative mammals has had significant impacts on native plants and animals in some areas of the boreal forest. In Finland, the North American white-tailed deer was introduced in the early part of the twentieth century and had become firmly established by 1934. American mink and muskrat were imported into northern and eastern Europe to be raised on farms for furs. Escaped muskrats have now spread across much of northern Eurasia. They have become a nuisance, especially because of their habitual burrowing into dams and dikes in the lowlands of north-central Europe. Raccoon dogs,

Sphagnum and Peat

Sphagnum is a genus that is composed of more than 150 species of mosses that grow in relatively cool moist environments where precipitation is common and evaporation rates are low. Quite commonly, sphagnum mosses are found growing at the surface in bogs, muskegs, mires, or peat bogs. As sphagnum grows as the top surface, the older moss materials beneath it die off. Decayed, partially dried, and compacted *sphagnum moss* is often referred to as *peat moss*. Sphagnum moss has a commercial use in the floral industry as a materials to line flower pots and wire baskets.

Peat is a material consisting of partially decomposed biological materials (primarily plants). The decomposition process is incomplete, primarily since the process occurs mostly when submerged in standing water without the benefit of free oxygen, leaving partially carbonized tissue that is capably of holding large volumes of water and which also can be severely compressed. The materials are used as soil conditioner, but also as a fuel when dried (home heating as well as electric power generation in Germany and Finland, for example).

native to southeastern Siberia, were intentionally released in European Russia between 1927 and 1957 to promote increased fur production. These fur-bearing animals spread westward; raccoon dogs reached Sweden by 1945. They had invaded most of Finland by the 1990s. The milder climates of the Baltic region do not prompt these animals to develop the long hairs they produce under colder climatic conditions. As a result, they have little or no value as fur-bearing animals in Sweden and the coastal regions of Finland. Instead, they have become a nuisance species, commonly feeding on native small game animals.

Predator control in North America is not that different from what has been done in northwestern Eurasia. Populations of the wolf and the brown bear have been eliminated or severely reduced throughout much of the boreal forest of North America. Large-scale surface mining operations (strip mining, exploration, and dredging), logging, and the construction of dams for hydroelectric power and flood control have all contributed to the fragmentation of the boreal forests and have altered or disrupted the dispersal routes of large mammals. During the construction of the Alaska Pipeline, the various aspects of such impacts on caribou were widely publicized. The recurring talk of expanding the oil drilling in Alaska brings these issues repeatedly to the forefront of environmental concerns. In the outliers of the boreal forests of North America, in the mountains, massive destruction of boreal forest habitats comes as a result of copper, coal, and iron mining; domestic live-stock grazing; and gas and oil exploration and development. This is especially evident in the Rocky Mountain region.

Throughout the past half-century North American boreal forests, especially those in public ownership, have been subjected to multiple-use concepts that often permit and encourage usages other than forestry and grazing: recreational use has increasingly contributed to the alteration or even destruction of forested areas. Ski resorts with their various runs and ski trails, associated residential developments, vacation home developments, shopping and entertainment, and the infrastructure support (transportation, utilities, waste water treatment, and so on) that is necessary for such developments all combine to have major impacts on boreal forest regions. A common and popular recreational activity is snowmobiling. It requires extensive trails throughout the southern regions of the boreal forest belt in North America. It is also a popular sport in the boreal forests of Scandinavia. Trails, noise, supply developments, and transportation access points have affected the boreal forests. Hunting does not by itself directly affect boreal forest trees. However, increasingly all-terrain vehicles, sport utility vehicles, and the like are used to access hunting areas and have pushed the inroads of recreational hunting far into the boreal forest.

Wherever large concentrations of people live in proximity to forested regions, urbanization typically affects downwind conifer forests through the effects of air pollution. The redwood forests are among the better-known and often-visited forests in the western United States. It is estimated that less than 4 percent of the original redwood forests remain. Their loss is attributed to a combination of past and present logging activities and the effects of urban development upwind (primarily on the Pacific Coast) of the redwoods. Particularly damaging to needleleaf trees is the ozone generated when the vehicular exhaust reacts with sunlight. Conifers are susceptible since their needles are exposed year-round. Scientist have found that the effects of ozone in the Sierra Nevada are especially evident in stands of white fir, ponderosa pine, and Jeffrey pine, as well as some of the lichens in the groundcover.

Global warming is altering the environmental conditions within boreal forest regions. Temperature changes will result in habitat losses. For much of the Canadian boreal forests the total habitat losses are predicted to be above 50 percent. Slight temperature increases will push the arctic treeline (the northern boundary of the Boreal Forest Biome) significantly northward to an extent that will seriously diminish the size of the Tundra Biome as well as reduce the aquatic ecosystems that are closely associated with both biomes.

The effects of logging, mining (including the past activities of hydraulic mining and dredging for gold), recreational activities, and urban development certainly have eliminated or greatly changed parts of the boreal forest. Those impacts may be minor and spatially limited in light of the widespread changes yet to come. Global climatic changes that have been ongoing for some decades, but are better documented in recent years, involve a slight warming trend in the climates of the Northern Hemisphere. The eventual consequences of climatic change are not really known and are still subject to a lot of speculation and debate. Most of the existing scientific models that attempt to predict the effects of global warming indicate that the impacts will be greatest in those latitudes currently occupied by the circumpolar boreal forests of the Northern Hemisphere. There is, however, no consensus on the eventual outcomes of such changes—for example, are they going to be regarded positively or negatively? Current studies reveal that a very slight increase in temperatures might increase the rate of nutrient cycling in the boreal forest regions. This could be regarded as economically beneficial, since it would increase the rate of growth of trees, thus increasing timber production. However, even greater increases in temperatures could lead to the destruction of the boreal forests. Since these forests are indeed a major source of timber and other wood products (primarily pulp), their demise would severely affect the economies of Russia, Canada, Finland, Norway, Sweden, and the United States. Boreal forest regions seem to be highly responsive to climatic change. Because of the potential economic impacts such changes might bring about, they are a major focus of study for researchers who investigate global change of climatic factors.

Further Readings

Bailey, R. G. 1994. *Ecoregions of the United States.* Washington, DC: U.S. Forest Service. http://www.fs.fed.us/land/pubs/ecoregions/ecoregions.html.

FAO. 2001. Global Ecological Zoning for the Global Forest Resources Assessment. http://www.fao.org/docrep/006/ad652e/ad652e00.htm.

Runesson, Ulf T. 2007. Boreal Forests: Overview. http://www.borealforest.org/index.php?category=world_boreal_forest&page=overview.

Appendix

Plants and Animals of the Boreal Forest Biome Mentioned in the Text (arranged geographically)

North America

Plants of the Boreal Forest Belt

Needleleaf trees

White spruce	*Picea glauca*
Black spruce	*Picea mariana*
Larch or Tamarack	*Larix larichina*
Jack pine	*Pinus banksiana*

Broadleaf deciduous trees

Paper birch	*Betula papyrifera*
Balsam poplar	*Populus balsamifera*
Black cottonwood	*Populus trichorarpa*
Willows	*Salix* spp.
American green alder	*Alnus crispa*
Sitka alder	*Alnus sinuata*

Shrubs

High bush cranberry	*Viburnum edule*
Labrador tea	*Ledum groenlandicum*
Narrow-leaf Labrador tea	*Ledum decumbens*
Mountain cranberry	*Vaccinium vitis-idaea*
Prickly rose	*Rosa acicularis*
Red-fruit bearberry	*Arctostaphylos rubra*

Mammals of the Boreal Forest Belt

Herbivores

Barren ground caribou	*Rangifer tarandus*
Woodland caribou	*Rangifer caribou*

Moose	*Alces alces*
Wapiti	*Cervus elaphus*
Snowshoe or Varying hare	*Lepus americanus*
Beaver	*Castor canadensis*
Muskrat	*Ondatra zibethica*
Porcupine	*Erethizon dorsatum*
Red squirrel	*Tamiasciurus hudsonicus*

Carnivores

Lynx	*Felis lynx*
Wolf	*Canis lupus*
Coyote	*Canis latrans*
Red fox	*Vulpes vulpes*
Brown or Grizzly bear	*Ursus arctos*
Pine marten	*Martes americana*
Fisher	*Martes pennanti*
Ermine	*Mustela erminea*
Mink	*Mustela vison*
Long-tailed weasel	*Mustela frenata*
Least weasel	*Mustela nivalis*
River otter	*Lutra canadensis*
Wolverine	*Gulo gulo*

Birds

Spruce Grouse	*Canachites canadensis*
Sharp-tailed Grouse	*Pediocetes phasianellus*
Willow Ptarmigan	*Lagopus lagopus*
Barred Owl	*Strix varia*
Great Gray Owl	*Strix nebulosa*
Boreal Owl	*Aegolus funerus*
Saw-whet Owl	*Aegolus acadius*
Black-backed Woodpecker	*Picoides articus*
Three-toed Woodpecker	*Picoides tridactyla*
Pileated Woodpecker	*Dryocopus pileatus*
Gray Jay or Whiskeyjack	*Perisoreus canadensis*
Black-capped chickadee	*Poecile atricapillus*
Boreal chickadee	*Poecile hudsonicus*
Hermit Thrush	*Catharus guttatus*
Swainson's Thrush	*Catharus ustulatus*
Bay-breasted Warbler	*Dendolimus castanea*
Cape May Warbler	*Dendrolimus tigrinia*
Palm Warbler	*Dendroica palmarum*
Red Crossbill	*Loxia curvirostra*
White-winged Crossbill	*Loxia leucoptera*

(Continued)

Redpoll	*Acanthus falmea*
Purple Finch	*Carpodacus purpureus*
White-throated Sparrow	*Zonotrichia albicollis*

Insect

| Spruce budworm | *Choristoreura fumiferana* |

Plants of the Pacific Northwest

Needleleaf trees

Sitka spruce	*Picea sitchensis*
Grand fir	*Abies grandis*
Pacific silver fir	*Abies procera*
Western hemlock	*Tsuga heterophylla*
Mountain hemlock	*Tsuga mertensiana*
Shore pine or Lodgepole pine	*Pinus contorta*
Alaska cedar	*Chamaecyparis nootkatensis*
Port Orford cedar	*Chamaecyparis lawsoniana*
Douglas fir	*Pseudotsuga menziesii*
Western red cedar	*Thuja plicata*
Redwood	*Sequoia sempervirens*

Broadleaf trees

Red alder	*Alnus rubra*
Sitka alder	*Alnus sinuata*
California laurel	*Umbellularia californica*

Shrubs

Barclay willow	*Salix barclayi*
Scouler willow	*Salix scouleriana*
Sitka willow	*Salix sitchensis*
Currants	*Ribes* spp.
Evergreen huckleberry	*Vaccinium ovatum*
High bush cranberry	*Viburnum edule*
Pacific rhododendron	*Rhododendron macrophyllum*
Salal	*Gaultheria shallon*
Rusty menziesia	*Menziesia ferruginea*

Forbs and subshrubs

Devilsclub	*Oplopanax horridus*
Evergreen violet	*Viola sempervirens*
Wood violet	*Viola glabella*
False lily-of-the-valley	*Maianthemum dilatatum*
Oregon oxalis	*Oxalis orgenana*
Smith's fairybells	*Disporum smithii*
Salmonberry	*Rubus spectabilis*

Mammals of the Pacific Northwest

Herbivores

Black-tailed mule deer	*Odocoileus hemionus*
Mountain beaver or Sewellel	*Aplodontia rufa*
Red squirrel	*Tamiasciurus hudsonicus*
Chickaree	*Tamiasciurus douglasi*
Townsend's chipmunk	*Eutamius townsendi*

Carnivores

River otter	*Lutra canadensis*
Bobcat	*Felis rufus*
Mountain lion	*Felis concolor*

Omnivores

Spotted skunk	*Spigale putrius*
Striped skunk	*Mephitis mephitis*

Birds

Spotted Owl	*Strix occidentalis*
Stellar's Jay	*Cyanocitta stelleri*
Chestnut-backed chickadee	*Poecile rufescens*
Varied Thrush	*Loxereus naevius*

Plants of the Eastern Coast Ranges and Western Cascade–Sierra Nevada Ranges: Western Hemlock Zone

Needleleaf trees

Sitka spruce	*Picea sitchensis*
Grand fir	*Abies grandis*
Shasta red fir	*Abies magnifica*
Western hemlock	*Tsuga heterophylla*
Douglas fir	*Pseudotsuga menzeisii*
Incense cedar	*Librocedrus decurrens*
Port Orford cedar	*Chamaecyparis lawsoniana*
Western red cedar	*Thuja plicata*
Western yew	*Taxus brevifolia*
Ponderosa pine	*Pinus ponderosa*
Sugar pine	*Pinus lambertiana*
Western white pine	*Pinus monticola*
Whitebark pine	*Pinus albicaulis*

(Continued)

Broadleaf deciduous trees
Red alder   *Alnus rubra*
Bigleaf maple   *Acer macrophyllum*
Golden chinkapin   *Castanopsis chrysophylla*

Shrubs
Pacific rhododendron   *Rhododendron macrophyllum*
Salal   *Gaultheria shallon*
Trailing blackberry or Thimbleberry   *Rubus parviflorus*
Oregon grape   *Berberis nervosa*

Forbs and other plants of the ground layer
Fireweed   *Epilobium angustifolium*
Pacific starflower   *Trientalis latifolia*
Woodland groundsel   *Senecio sylvaticus*
Whipple vine or Modesty   *Whipplea modesta*
Swordfern   *Polystichum munitum*

Plants of the Central Cascades and Sierra Nevada

Needleleaf trees
Englemann spruce   *Picea englemannii*
Grand fir   *Abies grandis*
Noble fir   *Abies procera*
Pacific silver fir   *Abies amabilis*
Subalpine fir   *Abies lasiocarpa*
Douglas fir   *Pseudotsuga menziesii*
Western hemlock   *Tsuga heterophylla*
Mountain hemlock   *Tsuga mertensiana*
Western larch   *Larix occudnetalis*
Jeffrey pine   *Pinus jeffreyi*
Lodgepole pine   *Pinus conorata*
Western white pine   *Pinus monticola*
Alaska cedar   *Chamaecyparis nootkatensis*
Giant sequoia   *Sequoia gigantea*
Western red cedar   *Thuja plicata*

Shrubs
Huckleberries   *Vaccinium* spp.
Pacific rhododendron   *Rhododendron macrophyllum*
Salal   *Gaultheria shallon*

Forbs and other plants of the ground layer
Dwarf blackberry   *Rubus lasiococcus*
Evergreen violet   *Viola sempervirens*

Plants of the Redwood Forest

Needleleaf trees
Redwood   *Sequoia sempervirens*
Douglas fir   *Pseudotsuga menziesii*

Broadleaf trees
Bigleaf maple   *Acer macrophyllum*
Tanoak   *Lithocarpus densiflorus*
California laurel   *Umbellularia californica*

Trees of the Rocky Mountains

Needleleaf trees
Englemann spruce   *Picea englemannii*
Subalpine fir   *Abies lasicarpa*
Douglas fir   *Pseudotsuga menziesii*
Bristlecone pine   *Pinus aristata*
Lodgepole pine   *Pinus contorta*
Ponderosa pine   *Pinus ponderosa*

Broadleaf deciduous tree
Quaking aspen   *Populus tremuloides*

Mammals of the Rocky Mountains

Herbivores
Bighorn sheep   *Ovis canadensis*
Moose   *Alces alces*
Wapiti or Elk   *Cervus elaphus*
Black-tailed mule deer   *Odocoileus hemionus*
Abert's squirrel   *Sciurus aberti*
Golden-mantled ground squirrel   *Citellus lateralis*
Least chipmunk   *Eutamias minimus*
Red squirrel   *Tamiasciurus hudsonicus*

Carnivores
Brown bear or Grizzly   *Ursus arctos*

Plants of the Appalachian Mountain Conifer Forests

Needleleaf trees
Red spruce   *Picea rubens*
Balsam fir   *Abies balsamea*
Fraser fir   *Abies fraseri*
Tamarack or American larch   *Larix laricina*

(Continued)

Broadleaf deciduous trees

Paper birch	*Betula papyrifera*
Yellow birch	*Betula alleghaniensis*

Mammals of the Appalachian Mountain Conifer Forests

Herbivores

Moose	*Alces alces*
White-tailed deer	*Odocoileus virginianus*
Snowshoe hare	*Lepus americanus*
Red squirrel	*Tamiasciurus hudsonicus*
Northern flying squirrel	*Glaucomys sabrinus*

Carnivores

Mountain lion	*Felis concolor*
Long-tailed weasel	*Mustela frenata*

Birds

Black-capped Chickadee	*Poecile atricapillus*
Common Raven	*Corvus corax*
Dark-eyed Junco	*Junco hyemalis*
Golden-crowned Kinglet	*Regulus satrapa*
Ruby-crowned Kinglet	*Regulus calendula*
Purple Finch	*Carpodacus purpureus*
Red-breasted Nuthatch	*Sitta canadensis*
White-throated Sparrow	*Zonotrichia albicollis*

Eurasia

Plants of the Fennoscandian Taiga

Needleleaf trees

Norway spruce	*Picea abies*
Scots pine	*Pinus sylvestris*
Juniper	*Juniperus communis*

Broadleaf trees

Birch	*Betula tortuosa*
Willows	*Salix* spp.

Shrubs

Bilberry	*Vaccinium myrillus*
Cranberry	*Vaccinium vitis-idaea*
Goat willow	*Salix caprea*
Rowan	*Sorbus aucuparia*

Mammals Animals of the Fennoscandian Taiga

Herbivores

Reindeer	*Rangifer tarandus*
Red deer (moose)	*Alces alces*
Roe deer	*Capreolus capreolus*
Mountain hare	*Lepus timidus*
European beaver	*Castor fiber*
Flying squirrel	*Pteromys volans*

Carnivores

Wolf	*Canis lupus*
Red fox	*Vulpes vulpes*
Lynx	*Felis lynx*
Brown bear	*Ursus arctos*
Wolverine	*Gulo gulo*
European mink	*Mustela lutreola*
American mink[a]	*Mustela vison*
European pine marten	*Martes martes*
Raccoon dog[a]	*Nyctereutes procyonoides*
North American muskrat[a]	*Ondatra zibethicus*

Note: [a]Introduced from North America.

Birds

Capercaillie	*Tetrao urogallus*
Black Grouse	*Tetrao tetrix*
Hazel Hen	*Tetrastes bonasia*
Siberian Spruce Grouse	*Falcipennis falcipennis*

Insect

Longhorn or Black fir sawyer beetle	*Monochamus urussovi*

Plants of the Taiga of European Russia

Needleleaf trees

Scots pine	*Pinus sylvestris*
Siberian cedar pine	*Pinus sibirica*
Siberian spruce	*Picea obovata*
Siberia fir	*Abies sibirica*
Siberian larch	*Larix sibirica*

Broadleaf trees

Birch	*Betula tortuosa*
Manchurian alder	*Alnus fruticosa*

(*Continued*)

Ground layer

Feather moss	*Pleurozium schreberi*
Lichens	*Cladonia* spp.

Plants of the Western Siberian Boreal Forest

Needleleaf trees

Siberian spruce	*Picea obovata*
Siberian cedar pine	*Pinus sibirica*
Siberian larch	*Larix sibirica*
Sukachev's larch	*Larix sukaczevii*

Ground layer

Feather moss	*Pleurozium schreberi*
Lichens	*Cladonia* spp.

Plants of the Central and Eastern Siberian Boreal Forest

Needleleaf trees

Dahurian larch	*Larix dahurica*
Siberian larch	*Larix sibirica*
Japanese stone pine	*Pinus pumila*
Siberian pine	*Pinus sibirica*

Ground layer

Feather moss	*Pleurozium schreberi*
Lichens	*Cladonia* spp.

Mammals of the Siberian Boreal Forests

Herbivores

Eurasian flying squirrel	*Pteromys volans*

Carnivores

Wolf	*Canis lupus*
Red fox	*Vulpes vulpes*
Lynx	*Felis lynx*
Brown bear	*Ursus arctos*
Wolverine	*Gulo gulo*
Eurasian river otter	*Lutra lutra*
Raccoon dog	*Nyctereutes procyonoides*
Sable	*Martes zibellina*

Birds

Capercaillie	*Tetrao urogallus*
Hazel Hen	*Tetrastes bonasia*
Siberian Spruce Grouse	*Falcipennis falcipennis*
Siberian Tit	*Parus cintus*
Willow Tit	*Parus montanus*

Insects

Longhorn or Black fir sawyer beetle	*Monochamus urussovi*
Siberian silkworm	*Dendrolimus superans*

3

·· ····

Temperate Broadleaf Deciduous Forest Biome

The Temperate Broadleaf Deciduous Forest Biome is primarily encountered in the humid middle latitudes of Earth, those that are characterized by temperate climatic conditions. These commonly have distinct annual seasons that are typified by a cool to cold period during the months of winter and by a warm to hot period during the summer. The dominant trees shed their leaves in the autumn and remain bare throughout the temperature-induced nongrowing winter season. Over several weeks in the spring, they produce new foliage and once again close the forest canopy. The biome is found nearly exclusively in the Northern Hemisphere. Only one important area of occurrence with comparable types of forests occurs in the temperate zones of the southern part of western South America (see Figure 3.1).

The temperate broadleaf deciduous forests of Europe and East Asia were decimated, converted to other land usages, or severely altered by the time the scientific age began. The exploration and colonization of the New World by some western European countries were responses to their search for alternative and additional resources, including commercial trees and other plants. Thomas Hariot, a member of England's initial colonization effort at Roanoke Island on the eastern coast of North America in 1585, was the first to study the natural history of the settlement site. Because he was the first to provide us with descriptions of the environment encountered by the early settlers, he became the first to describe plants and animals from the temperate broadleaf deciduous forest of America.

Natural historians were not the only ones searching for, inventorying, and describing plants and animals from outside of Europe. One of the traditional driving forces behind the search for new plants was the desire to discover and to bring

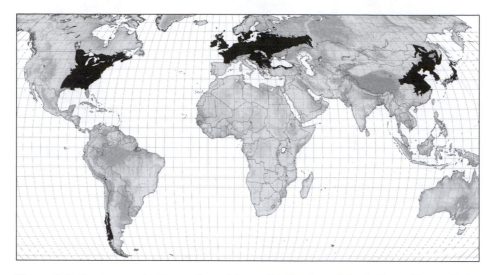

Figure 3.1 Temperate deciduous forest biome distribution. *(Map by Bernd Kuennecke.)*

back exotic plants to the formal gardens as well as Europe's new botanical gardens. Europe had wealthy patrons who could afford to finance expeditions to exotic places. Botanists were sponsored to travel the wild regions of eastern North America looking for new plants that would be suitable for transplant to Europe. In the mid-1700s, American John Bartram and, later, his son William were hired for such purposes. Francois Andre Michaux, a French botanist with European sponsors, searched for plants and animals in the New World, and in 1818 and 1819, he published the three-volume set *The North American Sylva, or a Description of Forest Types of the United States, Canada and Nova Scotia.*

More recently, E. Lucy Braun conducted an in-depth study of the eastern deciduous forests of North America. She developed and produced her by-now classic 1950 book, *Deciduous Forests of Eastern North America*, in which she set about classifying the forests into the nine types that are still recognized today. She attempted to develop an accurate picture and description of the forests of eastern North America as they existed prior to settlement by Europeans, and her book remains a standard reference on the biome in eastern North America. Ecological studies have become much more important since the 1950s. They have supplanted the previously popular descriptive or natural history works. The temperate broadleaf deciduous forests of eastern North America have been a test-bed of sorts for the development of pioneering studies on ecological succession of natural forest stands, energy flow in forests, and the nutrient cycles that are involved. One outcome has been the careful delineation of ecological regions. Much research continues today, especially in conjunction with issues related to global warming and human impacts on the forest ecology in eastern forests of the United States.

Global Overview of the Biome

One of the largest and still mostly intact temperate broadleaf deciduous forests constitutes the natural vegetation throughout much of the eastern United States and a small portion of southeastern Canada, where it forms a broad ecotone with the Boreal Forest Biome (see Plate IX). In the historic past, a second tremendously extensive cover of temperate broadleaf deciduous forests was the natural vegetation of western and central Europe. It extended from the northern fringes of the Iberian Peninsula and the British Isles eastward across the continent through most of western, central and north central, and eastern Europe to the Ural Mountains. The third major region of the biome in the Northern Hemisphere is in eastern Asia. Its approximate southern boundary lies north of the Yangtze River (Chang Jiang) at approximately 30° N. From its southern limits, the temperate forest extends northward across eastern China to latitudes of 50° N to 60° N in Manchuria, in the northeast of China, and across the easternmost sections of Russian Siberia as far as the Kamchatka Peninsula. The eastern edge occurs in central Japan. The western limits of this biome in East Asia are along a line roughly formed by the 125° E meridian in the northwest to longitude 115° E in the southwest of China.

In the Southern Hemisphere, the deciduous forests are composed of completely different trees than those encountered in the Northern Hemisphere. A minor, restricted biome in South America, the temperate broadleaf deciduous forest occurs on either side of the Andes in the westernmost part of Argentina and Chile approximately south of latitude 37° S. The dry grasslands of Patagonia in Argentina form the eastern boundary.

The three major expressions of the Temperate Broadleaf Deciduous Forest Biome within the Northern Hemisphere are geographically disjunctive. Significant distance isolates each from the others. In the geologic past, however, they were part of a continuous forest belt across North America and Eurasia. Biological evidence in the form of plants that are today still closely related supports such a statement. The most important and typical trees of the forests of this biome are members of the beech family (Fagaceae), which includes the beeches and the oaks. Trees from this family are found in all temperate deciduous forests of the Northern Hemisphere. Other important trees with a wide distribution are members of the birch family (Betulaceae), such as hazels, hornbeams, and hophornbeams; the walnut family (Juglandaceae), among them in North America the hickories in particular; and the maple family (Aceraceae). These forest trees, in various combinations and proportions, together with more localized species, are characteristic of the temperate broadleaf deciduous forest. The canopy of these forests is opened seasonally due to the deciduous habit of the majority of the trees, permitting sunlight to reach the forest floor during later fall, winter, and in the early part of the growing season. This allows many forbs and shrubs to grow and flower before the foliage of the tallest canopy trees is fully developed and shades the forest floor once again. Understory and groundcover plants can go through much of their annual cycle of

Figure 3.2 Deciduous forest with understory and groundcover. *(© Susan L. Woodward, used by permission.)*

flowering and beginning seed development before the lack of sunlight due to shade from the canopy prevents these essential reproductive stages. Thus, almost all of these forests have an understory of plants at the ground or shrub layer (see Figure 3.2).

The temperate broadleaf deciduous forests contain the most species-rich vegetation in the middle latitudes. Since much of this biome is located in hilly to mountainous areas, a high number of microclimatic variations, which are closely related to similarly high variations in soils, slopes, water moisture levels, and so on, contribute to a greatly heightened diversity of plants in different stands at various elevations. Due to the environmental conditions that change dramatically over short distances in mountainous terrain, stands with different plant combinations are often only a few feet away from sites with yet other combination of species.

The forests of East Asia are thought to be the oldest. And, in all likelihood, they were the least affected by the climatic changes of Pleistocene Epoch and its Ice Ages (see Figure 3.3). This is probably why forest stands in this region today exhibit the greatest diversity of trees anywhere in the biome. The flora includes representatives of ancient genera and of genera limited in their natural distribution to eastern Asia. Also in this region are genera shared only with western North America. The North American expression ranks second in tree diversity to the East Asian part of the biome. The Temperate Broadleaf Forest Biome of Europe has the lowest number of tree species.

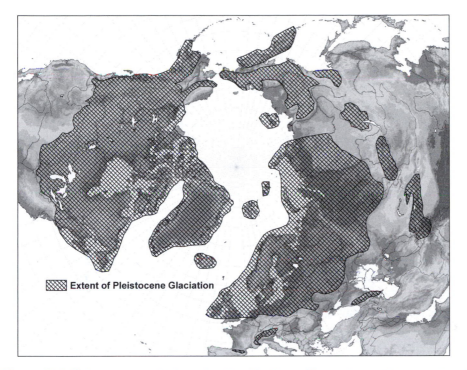

Figure 3.3 Pleistocene glaciations in the Northern Hemisphere. *(Map by Bernd Kuennecke.)*

The annual decomposition of nitrogen-rich foliage accumulating on the forest floor of temperate broadleaf deciduous forests contributes to relatively high natural fertility (at least in the A horizons) of the soils of these forests. The high annual accumulation of new humus materials also feeds the living organisms in these soils, from microorganisms to soil-aerating earthworms and small burrowing mammals. The deeply developed and noncompacted soils of these forests regions were, for a long time in the history of agriculture, the best soils available to farmers. It was relatively easy to remove the natural vegetation cover and to establish crops. For 2,000–3,000 years, agriculturalists in the temperate regions of western and central Europe repeatedly created clearings in forested areas and established farming villages. (The suffix "-loh" is attached to a multitude of place names across northern Germany, indicating that the village was a clearing in the forest at one time.) Only after the invention of the steel plow in the eighteenth century, were farmers able to successfully turn over the sod of temperate grasslands and expose even richer soils.

In East Asia, agricultural activities are intense and have been so for a long time in the history of humankind. As a direct result of such activities very little of the original natural forests of the region remain intact. The same is true of most regions of European deciduous forests. Only small "pockets" of natural forests are preserved (often because they were formerly feudal hunting preserves). However, even though the label "natural forest" or even "wilderness" is used at times to denote such small pockets of "natural old forests," these stands probably are not truly natural. The very nature of keeping such stands as game preserves required a purposeful selection of all plants, including the trees, permitted to remain on such sites.

Today, the largest tracts of this forest biome are found in eastern North America. Although extensive, most of the temperate broadleaf deciduous forests in North America are second growth forests. This is the result of past logging and the clearing of forests for agricultural lands, followed later by a conversion back to forests when agricultural areas were abandoned.

Climate

Throughout all areas of their occurrence, temperate broadleaf deciduous forests are associated with temperate climate conditions. These are expressed as humid subtropical (Cfa and Cfb in the Koeppen classification of climate types) (see Figure 3.4) and hot summer, humid continental climates (Dfa climate types) (see Figure 3.5; see also Table 1.1). The precipitation in areas of temperate deciduous forest is usually evenly distributed throughout the year. The average annual precipitation ranges from 30 to 50 in (750 mm to 1,250 mm), with localized higher rainfall where orographic effects exist. Winter seasons are commonly cool to cold and typically have at least some snowfall.

In all regions of the biome, summer seasons are warm to hot. In the mountains, of course, a vertical zonation of temperatures can result in cool summers. The growing season is typically at least six months long and can be much longer in the southern areas of the biome in North America, Western Europe, and East Asia, as well as in the northern areas of the biome in South America. Typically four distinct seasons are experienced: spring, summer, autumn, and winter. In eastern China, the influence of the Asian monsoons is significant, as is the influence of the great continental mass to the west, which gives the interior of this region, in particular, a near-continental climate. Summers are extremely humid and hot. Several cities in the drainage basin of the Yantgze River are referred to as the "furnaces of China." This region typically experiences rather dry and very cold winter seasons, but this is of little consequence to the vegetation since the dry season coincides with the time of year when most plants are dormant. Therefore, China has the same types of trees that thrive in the other Northern Hemisphere regions of the biome, where precipitation is year-round and temperatures not as extreme.

Geologic foundations

Two types of mountain system, each from a different geologic era, and various types of sedimentary covers of underlying structures dominate the physical structure of those regions of the Earth where the Temperate Broadleaf Deciduous Forest Biome occurs. Caledonian and Hercynian mountain roots (or Appalachian remnants) are

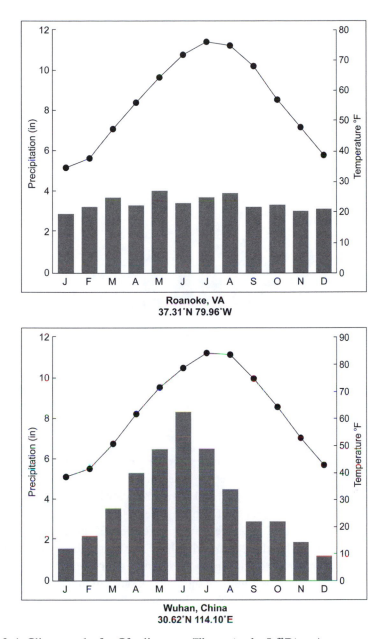

Figure 3.4 Climographs for Cfa climates. *(Illustration by Jeff Dixon.)*

found in North America, central Europe, and the British Isles. These are hilly to low-relief mountains originally formed during the Paleozoic and Mesozoic Eras (208–570 million years ago [mya]). They have experienced no mountain-building since then, but have been worn down by erosional processes to the mostly gently sloping landforms that we find today. The second type of mountain system is seen in

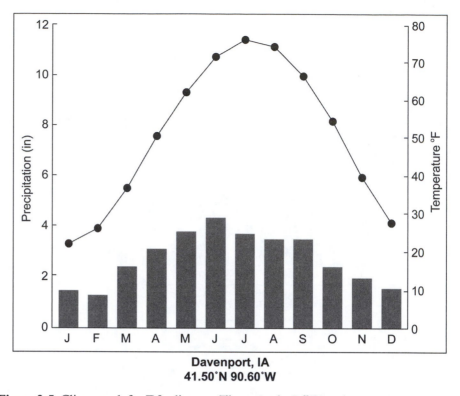

Figure 3.5 Climograph for Dfa climate. *(Illustration by Jeff Dixon.)*

the alpine systems of South America, East Asia, and west-central Asia. These mountains are relatively young, having been formed since the Jurassic Period (66–208 mya). They are typically more rugged and have a high local relief with moderate to steep slopes. Sedimentary covers are the third type of landform structure on which the temperate broadleaf deciduous forests have developed. These consist of sedimentary rock layers beyond shield exposures that have not been subjected to mountain-building forces. Sedimentary rock covers underlying structures of typically much greater age. Such structures are found on the edges of older mountain blocks, such as the Appalachian System, across the North European Plains to the Ural Mountains, and on the fringes of the Caucasus Mountains in western Asia.

Soils

Two types of soils are most common throughout the Temperate Broadleaf Deciduous Forest Biome, with a third encountered only occasionally. The soils are all products of podsolization soil-forming processes. Alfisols are moderately weathered forest soils and are one of the most widely distributed types of soils. These are

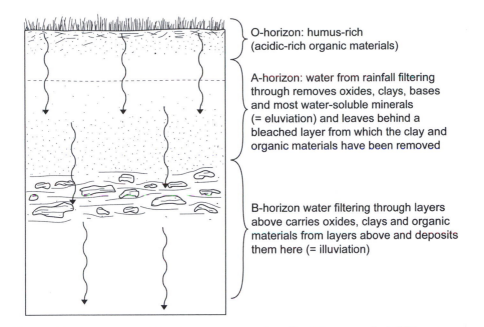

O-horizon: humus-rich
(acidic-rich organic materials)

A-horizon: water from rainfall filtering
through removes oxides, clays, bases
and most water-soluble minerals
(= eluviation) and leaves behind a
bleached layer from which the clay and
organic materials have been removed

B-horizon water filtering through layers
above carries oxides, clays and organic
materials from layers above and deposits
them here (= illuviation)

Figure 3.6 Eluviation and illuviation processes in soils. *(Illustration by Jeff Dixon.)*

gray forest soils of moist climates characterized by a B horizon that has been enriched (as a result of abundant water percolating through the soils) by accumulated silicate clay minerals (illuviation) that have a fair to moderate capacity to retain plant nutrients (base cations) such as magnesium and calcium (see Figure 3.6). These soils typically develop in the cooler areas of these forests. Udalfs are a subgroup of the Alfisols that typically develop under humid continental and hot summer climatic conditions.

Ultisols are highly weathered forest soils that have developed in warmer and moister conditions than the alfisols. (An alfisol may degenerate into an ultisol given time and exposure to an increased rate of weathering under moist conditions.) The increased precipitation in the ultisol regions results in a much greater rate of mineral alteration. This, in turn, means more leaching (eluviation), which leads to a lower level of base cations and hence to a lower level of natural soil fertility. The less-fertile ultisols are often referred to as "red and yellow forest soils." These soils have a tendency to be reddish or yellowish because of the residual iron and aluminum oxides in the A horizon from which many other minerals have been leached.

Throughout the forests, the organic layer on top of the soils typically consists of a rather nutrient-rich leaf litter that accumulates each autumn with the falling of the leaves. This layer begins to decompose with rising temperatures in late winter and early spring, producing a dark slick layer, the humus layer. The beginning of the decay of the leaf layer in early spring mineralizes the nutrients within these dead leaf materials. Precipitation then washes these nutrients into the soil to the root zone of plants just as the plants need fertilizer for a rapid growth spurt in early spring. As the moisture from precipitation filters through the humus layer, humus particles are washed downward (that is, leaching) through the upper parts of the A horizon, staining the soil horizons below dark brown. Since the humus under broadleaf deciduous forests is significantly less acidic than that which develops under the needleleaf canopy of boreal forests, leaching of the A horizon in alfisols is far less severe than in spodosols. Iron and aluminum are not easily removed, but remain in the A horizon, together with silica compounds. It is the color of the iron and aluminum oxides (reddish brown or rust color for iron and yellowish for aluminum) that makes these "red and yellow forest soils." The chemical symbols of the elements aluminum and iron (Al and Fe, respectively) have given the soil order the name alfisols.

The nutrients leached from the decomposing leaf and other biological materials in the forest litter are typically carried through the A horizon and then accumulated in the B horizon, together with the dark brown humus that provides the Bhorizon with its distinct brownish color. The B horizon is of high importance to the growth of plants, since the roots of trees reach to this depth to access the mineral nutrients throughout the growing season. It is this same B horizon that farmers try to reach by plowing and turning and bringing it to the surface where the shorter root systems of cultivated crops can reach its nutrient supply.

Vegetation

The forests of the Temperate Broadleaf Deciduous Forest Biome have a multilayered structure from the canopy to the ground layer (see Figure 3.7). A closed canopy layer is formed by the tallest trees. Beneath the canopy layer is commonly a

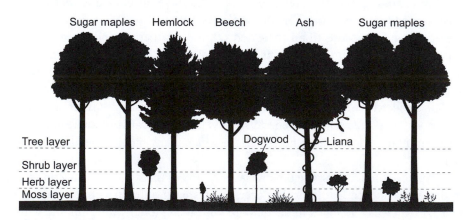

Figure 3.7 Schematic of layers in beech-maple-hemlock forests of the Smoky Mountains, North Carolina. *(Illustration by Jeff Dixon.)*

secondary layer consisting of the crowns of small trees or saplings. The trees of this understory layer do not reach the forest canopy even when fully mature. Typically a shrub layer forms the third layer. It is often made up of broadleaf deciduous and at times broadleaf evergreen plants. A rich herb layer of perennial forbs is common, as is a ground layer consisting of clubmosses, lichens, and mosses. Some growthforms encountered in the temperate broadleaf deciduous forests do not fit neatly into the normal categories of layers that are found, namely, woody vines (lianas). These are represented in the forests of North America by such plants as poison ivy and wild grape.

All life processes in the temperate broadleaf deciduous forests have a strong seasonal aspect to them, one of the key characteristics of this biome. All of the plants undergo seasonally distinct periods of vegetative growth and associated activities (development of foliage and growth of plants, flowering, and seed development), and they all have a dormant period during the coldest months of the year.

Structural changes within the leaves during the autumn season when the leaves change their colors precede the actual falling of the leaves. Leaves do not simply "fall" off. The first step is the development of an abscission zone at the base of the leaf petiole. Once this abscission zone has

Autumn Leaves

Perhaps the most visually obvious characteristic occurs annually as part of the seasonal changes: the changing color of leaves from green to yellows, reds, oranges, and brown during the respective autumn season. The color is brought about by biophysical processes in the plants. When the days grow shorter (less light) and the nights experience temperatures dropping off into the cooler ranges, the green chlorophyll of the foliage of broadleaf deciduous plants begins to break down. As a result, yellow (carotinoid) and red (anthocyanin) pigments in the leaves are exposed and provide the distinctive fall colors of these forests. The color intensity is not always the same from one year to another. It varies considerably depending on a number of conditions, such as moisture, moisture stress during the growing season, length of growing season during a particular year, onset of first frost, and so on. In those "good" years when the autumn foliage is particularly colorful and remains on the trees so that most trees "display" their fall colors simultaneously, these forests attract tourists in large numbers.

developed, a protective material forms over the surface, which becomes exposed when the leaf actually falls off and prevents the loss of water. During the dormancy of the winter season, this protective coating continues to prevent drying of the plant tissues.

With warming temperatures and longer days in spring, renewed activity begins as sap rises from the roots to the branches, well before the leaf buds actually open. Hazels and maples flower before their leaves begin to develop, though most trees wait until their leaves are beginning to form. As the initial leaf formation commences on these trees during the spring season, they go through the flowering and initial seed development stages. Basswoods (in Europe, called lindens or limes) are among the few trees that do not flower until after their foliage is fully developed in summer. The development of leaves brings about other activities, including some connected to insects. The fresh young leaves of spring typically contain the highest amount of nitrogen, phosphorus, and potassium compounds. Many insects thus coordinate the hatching of their larvae to this season, when there is an abundance of nutrients available in the newly developed foliage. A secondary effect of the timing of insects hatching involves the arrival of migratory birds. They seem to time their migrations to the explosion of the insect population witnessed each spring with the development of new leaves.

Throughout the early part of the growing season, the enrichment of the foliage with nutrients continues, albeit at a rate reduced from the initial leaf development stages. As the summer season unfolds, some of the plant nutrients are stored by the plants in the bark of twigs and stems. Other portions of the nutrients will seep out of the leaf cells and will be washed off by rainfall. During this time, magnesium and calcium compounds also are being accumulated in the leaves, where they are stored. When the leaves fall in autumn, these compounds will become part of the forest litter. The annual accumulation of leaves on the forest floor results in the development of thick humus layers in the temperate broadleaf deciduous forests. As the biomass on the forest floor decays, the nutrients are set free and can be washed into the soil horizons. This, in turn, maintains the nutrient-rich soils that the broadleaf trees require.

The annual cycle of foliage development, nutrient storage, leaves falling, decomposition of leaves, and enrichment of soils with nutrients maintains a tight nutrient cycle. A significant amount of energy is expended by trees in the annual production of new foliage, so the deciduous habit is only possible under environmental conditions that permit a sufficiently long growing season. The production of deciduous foliage is thus a successful alternative to producing well-protected evergreen leaves in moist temperate climates, as is shown by the overwhelming dominance of deciduous broadleaf species in the humid mid-latitude geographic regions, where these favorable environmental conditions are most often found.

The annual "greening-up" of the forest of the Temperate Broadleaf Deciduous Biome begins at the forest floor with the early development of the flowering perennial forbs. Their leaf developments, flowering, and seed development typically provides the first fresh green in these forests after a long winter dormancy. Throughout the North American temperate broadleaf deciduous forests, early growing and blooming species include wildflowers such as mayapple, spring beauty, and hepatica. Among the earliest to bloom are species such as skunk cabbage, which can bloom while snow is still covering the ground. The heat generated by the flower actually melts the overlying snow (provided the snow cover is not too thick). At the time of the year when snowmelt has occurred, but before the leaves of trees begin to enlarge, sunlight reaches the forest floor and warms it. This is when perennial shrubs take rapid advantage of a short period of time to complete their growth before trees begin their greening-up process. The shrub layer is typically the second layer of vegetation to respond to the early spring season. In the eastern United States, spicebush is the first of the shrubs to begin its development of leaves and blossoms, closely followed by other shrubs whose flowers are conspicuous in forest stands with still largely dormant trees, such as the highly visible blossoms of the redbud, shadbush or serviceberry, and flowering dogwood. All of these early developing plants in the ground and shrub layers experience a rapid slowing of their activities with the advent of the development of leaves by the trees of the sapling and canopy layers. Once the overstory trees have fully developed foliage, the lower layers will only continue their seed development, often over several months.

Animal Life

The Temperate Broadleaf Deciduous Forest Biome spans a great diversity of environmental conditions that sustains a tremendous variety of animal life. Invertebrates and vertebrates alike find these forests to be great providers of a variety of food resources and habitats. The food chain is powered to a large measure by the nutrient-rich plant matter that exists at various levels within this forest biome: the leaves, sap, and seeds; the nuts and berries; the rich nutrients in the bark of branches, stems, and trunks; the decomposing forest leaf litter and the dead and dying wood materials; the decay of dead animals; and so forth. The caterpillars of numerous butterflies, moths, and other Lepidoptera, such as bagworms, leaf rollers, and tentmakers, are all abundant consumers of green nutrient-rich foliage. Some simply chew the leaves, beginning on one edge and leaving behind partially eaten leaves. Others eat all leaf materials except for the veins, often leaving beautiful "skeletons" of leaves. Still others concentrate on devouring only the inner cells of leaves, leaving behind features that look like artistic scissor-cuts. A great variety of beetles (order Coleoptera) as well as flies (order Diptera) live on trees and in the forest litter. They feed on living as well as on decaying cell materials so abundantly encountered in these environments.

In the moister sections of the biome, vertebrates are represented by many kinds of amphibians, such as frogs, toads, tree frogs, and salamanders. A diversity of reptiles also exists and includes snakes, lizards, and turtles. The great variety and abundance of insects constitutes a food resource exploited by many insect-eating birds. Woodpeckers quite audibly drill and chip into the bark and decaying branches and trunks of trees to find grubs and beetles. Flycatchers typically pursue insects that move through the canopy of the trees, exhibiting (often acrobatically) flight patterns in this chase. Swallows normally feed above the canopy on insects that emerge from the treetops. Various members of the tit family actively search on leaves for insects. And on the ground, thrushes forage in the leaf litter for insects and larvae.

Rodents, such as voles, mice, and squirrels, as well as bats and insectivores (for example, shrews) are the most diverse groups of mammals in temperate broadleaf deciduous forests. Foxes are among the small predators typical of these forests. The populations of large mammals have been significantly altered by human influences. In some areas, large mammals have been eliminated by land use changes induced primarily through agricultural activities and the conversion of forests to residential use. Overhunting drastically reduced large mammals in some regions;

in others, deliberate campaigns to exterminate some species have been successful. The forests of eastern North America have experienced a near elimination of wolves and cougars, although wolves maintain a low population level in the northernmost parts of the biome. Cougars are rare, but reports of their sightings persist from Florida to Canada. Wood bison and elk have been eliminated from the forests in the southern and central regions of eastern North America. (Current efforts are under way in some areas of West Virginia to reintroduce elk into the temperate broadleaf deciduous forests.) In the Bialowieza Forest in Poland, a remnant population of the native wisent, which is a close relative of the North American bison, survives. (This population is actually a reintroduction and not part of the original herd). Several decades ago the forest-dwelling tarpan became extinct in the wild. It is a close relative of the domestic horse.

Seasonal rhythms influence the activity patterns of most vertebrates in these forests. The most common adaptation is the timing of reproduction to coincide with the greatest abundance of food. The winter hibernation of some animals is an adaptation to cold temperatures. The migration of some animals to regions with warmer temperatures is an adaptation to the scarcity of food supply during the winter season.

Rock crevices, animal burrows, caves, hollow logs, and even deserted anthills are used by snakes as places in which to hibernate. Frogs dig themselves into the mud and debris at the bottoms of ponds and lakes to hibernate. Salamanders and toads often seek protection from cold and desiccation and hibernate in caves, rotting logs, cracks and crevices in the soil, and even deserted rodent burrows. In the northern and central sections of the temperate broadleaf deciduous forests of eastern North America, about 75 percent of the summer-resident birds migrate south during the winter season. Those staying behind are typically omnivores, such as chickadees, or insectivorous birds, such as nuthatches and woodpeckers, both of which have the ability to pry dormant insects from beneath the tree bark. Most mammals do not migrate during the winter, since conditions are not sufficiently severe to force them into migration. However, bats in the northern sections of the biome do indeed migrate south during the onset of winter and hibernate in the more southern regions in natural tree cavities, buildings, or caves. Some native mammals in those regions of the biomes that experience severe winter weather do hibernate. Among these few are woodchucks, black bear, and eastern chipmunks, all of which disappear from the landscape during the coldest months.

MAJOR REGIONAL EXPRESSIONS OF THE TEMPERATE BROADLEAF DECIDUOUS FOREST BIOME

In the Northern Hemisphere, we encounter the biome in three distinctively different geographic regions (see Figure 3.1). They all share climatic and other physical environmental similarities with each other. The Southern Hemisphere has only one comparatively small area with an expression of this biome.

East-Central and Eastern United States and Southeastern Canada

The Temperate Broadleaf Deciduous Forest Biome in the eastern half of North America is in a geographic region bordered roughly by the 95° W meridian on the western side, latitude 48° N on the northern side, 30° N along the southern edge, and the Atlantic Ocean (see Figure 3.8). It corresponds roughly to the "temperate continental forest" and the "subtropical humid forest" in the global ecological zones outlined by the Food and Agriculture Organization of the United Nations (FAO). Along its northern edge, this biome borders the boreal forests. The approximate centerline of the broad ecotone of the two biomes reaches from the southern shores of Lake of the Woods (extreme southeastern corner of the Canadian province of Manitoba) to Thunder Bay on the western shore of Lake Superior. From there it stretches across the northern edge of the Upper Peninsula of Michigan and through southeastern Ontario, continuing on to the northern shores of the Gaspé Peninsula of Canada's Quebec province. The extensive ecotone region is one where we find great variations in local environmental conditions—wetlands, sand dunes, lakes, ponds, bogs, exposed crystalline rock with interspersed thin-soiled areas, old outwash plains

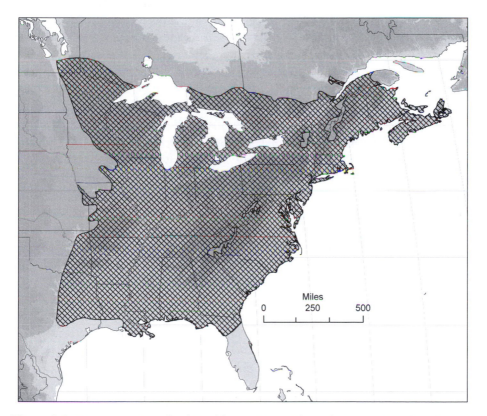

Figure 3.8 Temperate Broadleaf Deciduous Forest Biome in North America. *(Map by Bernd Kuennecke.)*

···

A Lost Tree

The American chestnut once was the largest tree in the eastern forests and was overwhelmingly dominant in some stands. The 1930s saw the demise of this tree throughout all areas where it used to be common and dominant. New York City experienced in 1906 the first attack of the fungus (*Cryphonectria parasitica*) that causes Chestnut blight. There is no certain proof but scientists today think that the introduction of the fungus happened accidentally when Chinese chestnut trees were imported as ornamentals. The spores of the fungus were carried by the winds, and the fungus rapidly spread throughout the Appalachian region west into Ohio and south to North Carolina. The fungus girdles and kills mature trees, but not seedlings or immature trees. American chestnut has the ability to resprout from its roots. As a result, some chestnut trees are still found in the shrub layer. However, as soon as these trees reach a size and an age at which they might reproduce, they become without fail infected by the blight and are killed. Individual trees of American chestnut that seem to be disease-resistant have been found, for example, in Maryland and Michigan. Efforts are under way to propagate such trees and use the seedlings to repopulate the temperate deciduous forests with its former proportion of American chestnut trees. It remains to be seen whether seedlings from these trees will indeed reach maturity without being overcome by this fungus.

···

and recent floodplains, microclimatic changes brought about by lake effects, mountains, and so on—and all conditions are interspersed with each other. This ecotone is a patchwork of forest stands belonging to either one biome or the other and contains mixtures of the two with various proportions of species from both.

The southern margins of the biome lie in the southeast United States, in the Coastal Plains, where warmer climatic conditions with higher precipitation levels prevail. One result is that soils have relatively low natural fertility. The broadleaf deciduous forests thus give way to the needleleaf and evergreen broadleaf trees more tolerant of low-nutrient soils. In the higher elevations of the Appalachian Mountains, the biome grades into an ecotone with boreal species before giving way to the boreal forests that form habitat islands along the spine of these mountains (see Chapter 2).

The biome does cover a large segment of North America and, as a result, spans a number of physiographic provinces. Each has its particular characteristics in terms of geology and geomorphology, climate and microclimate, and, of course, vegetation composition. The patchwork of forest stands throughout the biome that is created by different combinations of deciduous trees responding to local conditions is a reflection of the diversity of physical environments on the southeastern part of the continent. In her comprehensive study of the deciduous forests of North America, Braun identified a total of nine different forest regions making up the regional mosaic. These mosaics are the basis of the following descriptions.

Oaks, American beech, hickories, sugar maple, and basswoods are among the trees commonly encountered over much of the biome and are indicators of these different temperate deciduous forests today. Another species, the American chestnut, used to be one of the commonly encountered trees of these forests. However, it is now virtually extinct as a member of the canopy layer. Several evergreen conifers are commonly a part of the mix of trees in the temperate broadleaf deciduous forests. These include various pines and eastern hemlock. Both appear when local environmental conditions favor their development.

The oak-hickory-pine forest region identified by Braun coincides rather well with the Piedmont Plateau physiographic province. This region extends on the eastern side of the Appalachian Mountains from Pennsylvania to Alabama. It is a region in which the elevation ranges from 200 to 1,000 ft (60 to 300 m) above sea level. Surfaces are usually gently sloping, and high local relief is uncommon. It typically has hot summers with high humidity and cool (southern parts) to cold (northern parts) winters. The growing season extends from about 180 days in the north to 240 days in the south. Annual precipitation in the region varies from 42 to 52 in (1,090 to 1,345 mm). The soils in the oak-hickory-pine forest region are typically alfisols in the northern sections and more deeply weathered ultisols in the southern parts. Especially where ultisols occur, clay subsoils result from the illuviation of clay materials in the B horizon. The region's indicator species are numerous oak. Quite a few have adapted to fire and are able to resprout after a serious burn has seemingly killed the tree trunk. Other indicators are hickories; 14 of the 15 North American hickories are encountered within this forest type. Evergreen broadleaf trees are generally absent; however, evergreen needleleaf trees are common and constitute a major component of this forest region. Shortleaf pine and loblolly pine are common throughout the southern sections of this forest, while Virginia pine is more frequent in the northern sections. Flowering dogwood, sweetgum, sourwood, red maple, and black tupelo all are typical understory trees and are highly visible during their flowering season in early spring.

The majority of the so-called oak-chestnut forest region is in the Valley and Ridge and the Blue Ridge physiographic province. Braun's name is no longer suitable, since American chestnut is all

The New Jersey Pine Barrens

The New Jersey Pine Barrens, with nearly 1.1 million acres, are the largest and perhaps best known of several regions of varying sizes along the Atlantic Coast from New Jersey to Ontario that are commonly referred to as Pine Barrens. They consist of undulating sandy plains with low relief, sandy to gravely soils, and sluggishly flowing streams due to a lack of gradient to the sea. All have bogs and swamps in the many low depressions that mark the former beds of their meandering streams.

A specialized plant association occurs as the Pine Barrens of New Jersey and occupies the greater part of the coastal plan of that state. Scientists consider it part of the oak-hickory-pine forest region because it is, for the majority of the region, a forest of stunted pitch pine and shrubby oaks. Pitch pine is the characteristic tree of this region, occasionally replaced in importance by shortleaf pine, but always accompanied by various oaks such as blackjack oak, bear oak, and dwarf chinkapin oak. The understory often consists of shrubby versions of the oaks (bear oak and dwarf chinkapin oak), accompanied by heaths, such as mountain laurel, blue huckleberry, and box sandmyrtle. Interspersed between the old (unaffected by Pleistocene glaciation) sand and gravel deposits that make up a large portion of the Pine Barrens are patches of swampy terrain caused by poor drainage in the low-lying areas. Here the soil moisture is too high for pines and instead Atlantic white cedar, red maple, and black tupelo grow. The interspersed bogs and swamps support cranberries.

but extinct. It is better to simply call this the Appalachian oak forest region. This part of the Appalachian Mountains has elevations ranging from 1,000 to slightly more than 6,500 ft (300 to 2,000 m). High relief may be encountered locally. Considerable variation in annual rainfall occurs over the region in which precipitation

Temperate Forest Biomes

ranges from 30 to 70 in (760 to 1,780 mm). Some mountain chains within the Appalachian system cause orographic precipitation on one side and a drier rain-shadow on the other. Maximum rainfall occurs in the Great Smoky Mountains, at the southern end of the Blue Ridge, where an annual total of 90 in (2,300 mm) has been recorded. Within this forest region, the greatest number of forest communities occurs in the Southern Appalachians, an area of many mixed stands, typically dominated by white oak and chestnut oak. Normally other oaks, such as northern red oak, black oak, and scarlet oak, accompany the dominants in these mixed stands (see Figure 3.9). On moist lower slopes, the tall, straight tulip poplar is common.

Eastern hemlock frequently grows in cool ravines, especially on north- and east-facing slopes. Throughout the region, various sheltered valleys, known locally as coves, contain species-rich forest communities that closely resemble mixed mesophytic forests (described below). The cove forests can be distinguished from the surrounding stands of Appalachian Oak forest by the addition of other species into the mix, including varying proportions of American beech, tulip poplar, eastern hemlock, sugar maple, yellow buckeye, yellow birch, white basswood, and silverbells. Within the Appalachian oak forest region, with increasing elevation, generally above 4,500 ft (1,400 m), the less cold-tolerant species drop out of the mix (see Figure 3.7). The stands may then be referred to as "northern hardwoods," and

Figure 3.9 Oak-hickory-poplar forest, Appalachian Mountains, Virginia. *(© by L. Sue Perry, by permission.)*

include yellow buckeye, sugar maple, yellow birch, and American beech. Typically, the understory of small trees such as mountain maple, striped maple, and maple-leaved viburnum, is well developed. The ground layer normally contains a great variety of spring annual forbs. The wildflower display in early spring, before the dominant deciduous trees put forth their leaves, can be truly spectacular. One common wildflower is the mayapple (see Figure 3.10). The mix of trees at the higher elevations (above 5,000 ft or 1,520 m) changes rapidly from broadleaf deciduous trees to what is essentially a boreal forest of red spruce or Fraser fir (see Chapter 2). Mountain ash and yellow birch may grow on sites that are somewhat sheltered from the cold winter winds.

The mixed mesophytic forest region occupies the Cumberland Plateau, which adjoins the western boundary of the Southern Appalachians. It is that part of the Appalachian Plateau physiographic province that remained unglaciated during the Pleistocene Epoch. The region consists of an undulating surface of closely spaced hills and hollows. Fluvial processes, fed by ample precipitation, have dissected this landscape into one with few flat areas in either river valleys or hilltops. It is, for the most part, a continuously sloping surface. The southern location affords this region a long growing season with relatively mild winters combined with ample precipitation well distributed throughout the year. One of the consequences is that soils in this region are deeply weathered and show signs of leached A horizons and enriched B horizons. The vegetation mosaic of the region is referred to as "mixed mesophytic."

Figure 3.10 Mayapple in the Appalachian oak forest region. (© *by L. Sue Perry, by permission.*)

The term derives from the fact there is a significant overlap of plants with a generally more northerly distribution with those that have a more southerly distribution, as well as those that are typically encountered as dominants farther to the west. Added to this mix are plants that are largely restricted to the Southern Appalachians in general or to the Cumberland Plateau specifically. As a result of such overlaps in species distributions, this forest type has the greatest diversity of trees in the North American Temperate Broadleaf Deciduous Forest Biome. About 33 different tree species reach the canopy layer. Several trees generally share dominance in any given stand. Nine deciduous trees are dominant in virtually all plant communities. These include northern red oak, tulip poplar, sugar maple, white basswood, black tupelo, black walnut, cucumber tree, white ash, and—until a few decades ago—American chestnut. In addition to these dominants, numerous smaller trees occupy the sapling layer, trees like striped maple, sourwood, and the deciduous magnolias such as umbrella magnolia, Fraser magnolia, and the bigleaf magnolia. Not quite as common in the sapling layer and more localized in their distribution are flowering dogwood, redbud (mostly in areas that have been recently disturbed), pawpaw, American hornbeam, and eastern hophornbeam. Corresponding to the diversity of trees in the canopy and sapling layers, the shrub and herb layers are equally diverse throughout this region.

The southeastern evergreen forest has the greatest geographic extent of the temperate broadleaf deciduous forest types of Eastern North America. The forest extends from the Gulf-Atlantic Coastal Plains from southern Texas to the mouth of the James River in Virginia. Braun identified it as one of the nine temperate broadleaf deciduous forest types. A closer examination of the vegetation reveals, however, that broadleaf deciduous trees are mostly restricted to bottomlands or floodplains in river valleys, in and along the edges of swamps, and in "pocket" areas, such as the steep ravines sheltered from the wildfires that frequently occur naturally throughout this region. On the better-drained—although heavily leached and nutrient-poor—soils, trees such as southern red oak, turkey oak, blackjack oak, post oak, and sweetgum grow. In swamp forest stands, a deciduous conifer tree (the bald cypress) commonly grows alongside deciduous broadleaf trees such as water tupelo and Ogeechee tupelo, willows, and sycamore. Sycamores are found wherever ample water is available to its root system. The descriptions of early explorers provide us with some information about the natural vegetation cover of much of this region prior to the settlement by Europeans. The land areas between river valleys were most likely dominated by fire-resistant longleaf pine. Most stands of longleaf pine have been removed in conjunction with clearing for settlement and as a result of conversion to managed forests. Replanting of abandoned farm lands to forests, as well as modern tree farm practices, favor the conversion of the region to forests consisting mostly (and often solely) of slash pine and loblolly pine. Where evergreen broadleaf trees characterize the vegetation, stands are dominated by only a few species, among them live oak, southern magnolia, and American holly. Within the fringes of the

Appalachian Mountains and on the Coastal Plain, upland shrub-bogs or pocosins accompany some areas of poor drainage. These are typically dominated by evergreen shrubs, mostly members of the heath family.

The western mesophytic forest region reaches from the western edges of the Cumberland and Allegheny Plateaus to the eastern boundary of the Mississippi alluvial plain, and from northern Mississippi and Alabama northward to the southern limits of the former glaciation (see Figure 3.3) in eastern Indiana and Ohio. It encompasses the area of the Ozark Hills in Illinois. It is a broad transition zone from the mixed mesophytic forest to the prairies farther west. The forests are not uniform throughout this region, although northern red oak, white oak, and black oak, as well as several hickories and their plant associations seem to form a complex vegetation mosaic. In general, the region receives less precipitation than regions to the east, and it gradually becomes drier toward the west. Coinciding with the gradient of decreasing precipitation is an increasing occurrence of oaks. This forest region is much less diverse in its plant communities than the mixed mesophytic forest farther to the east. Terrain characteristics, microclimatic differences, and soil characteristics vary throughout and have been used to delineate six different sections of the mixed mesophytic forest region.

The Bluegrass section is frequently characterized by blue ash and bur oak, while other oaks, ash, maple, and hickory are within the association. Undergrowth is largely absent in this section. The Nashville Basin is unique in that such areas underlain by dolomite and limestone have forest communities of localized importance typically referred to as cedar glades. The dominant species is red cedar. These groves typically have a healthy mix of post oak and chinkapin oak, shagbark hickory, and winged elm. In the sapling layer, redbud grows—a fact that becomes quite evident when this tree blooms before the leafing-out of other trees in the stands.

An oak-hickory forest commonly dominates the western sections of the biome throughout the Central Lowlands and Interior Highlands physiographic provinces (see Figure 3.8). Farther north, across the formerly glaciated plains of Indiana and Ohio, is the beech-sugar maple forest region. Although the name implies that both are dominant trees of the canopy, it is indeed the beech that is most abundant. Sugar maple typically dominates the understory. North of this forest region is the so-called Driftless Area of Wisconsin, a geographic area that was bypassed by the continental glaciers of the Pleistocene. This relatively small, poorly delineated area in the northernmost part of the biome is the sugar maple-basswood forest region.

The ninth and final forest region that was outlined by Braun in 1950 as the northern hardwoods-conifer region lies in glaciated areas of the northeastern United States, including parts of New York and New England. Here forest stands exhibit mixtures of coniferous trees belonging to the boreal forests to the north, such as white pine and other pines, red spruce, balsam fir, and eastern hemlock, that associate with northern hardwoods, such as sugar maple, American beech, and northern red oak.

Lungless Salamanders

Lungless salamanders have undergone some interesting evolutionary stages. The woodland group (genus *Plethodon*) has dispensed entirely with an aquatic larval stage. Instead, eggs are laid and the young develop under logs and rocks in the forest, often significant distances away from the mountain streams in which they most likely evolved. Some lungless salamanders are distributed throughout the southern Appalachians. Others, such as the Peaks of Otter salamander, have extremely small distribution areas. The red-backed salamander has a distribution that ranges throughout much of the temperate broadleaf deciduous forest. It is one of the most abundant vertebrates found in this biome.

Animal life

The great environmental diversity of the temperate broadleaf deciduous forest region of Eastern North America accounts not only for the multitude of plant associations, but also for the significant variety in the wildlife that resides either permanently or seasonally in these forests. This diversity is enhanced by the presence of various successional stages in the vegetation associations. The amphibians are well represented in terms of numbers of species. Biologists have noted, in particular, the lungless salamanders found in this biome. They reach their highest level of diversity in the southern Appalachian Mountains, where some 30 different kinds are endemic to the region.

Lizards are common but not as widespread as salamanders. Biologists have enumerated some 11 different kinds of lizards from the deciduous forest region. Among these, the eastern fence lizard is by far the most widespread. Snakes are relatively common throughout the biome. Most are members of the family Colubridae, including the common garter snake, the common kingsnake, the racer, and the rat snake. Among the poisonous snakes of the region, perhaps the best known is the timber rattlesnake, a venomous viper (family Viperidae). Nearly as frequently encountered is the copperhead. Both may be found more or less commonly throughout the biome depending on local environmental conditions.

The biome's high diversity is also mirrored by the numbers and varieties of birds that are either resident or seasonal in the temperate broadleaf deciduous forests. As mentioned above, a large proportion of the birds are migratory insect-eaters. They arrive at a time when insect populations experience their seasonal explosions in early spring, and they leave with the onset of the colder season and its greatly diminished supply of insects. These migratory birds as a group are referred to as Neotropical migrants. Characteristically, they winter in tropical environments south of the United States, in Central and South America. One of the most common of these songbirds is the Red-eyed Vireo. Highly vocal, it typically gathers insects from the surface of leaves in these forests. The wood warbler group constitutes a large portion of the migrants. Wood warblers are observed wherever it is moist and there is a well-developed understory of densely growing saplings and shrubs. The Black-and-white Warbler and Chestnut-sided Warbler are common in the northern regions of the biome. Throughout the southern sections, the Yellow-throated Warbler and the Hooded Warbler are the more frequently seen and heard warblers. While warblers are perhaps the most commonly noticed migratory songbirds, numerous other songbirds come to these forests during the

Decline of Neotropical Migrants

Neotropical migrants are those birds in North American forests that come north to breed because of the bountiful food supply in these forests during the summer season (insects and seeds), which enables them to raise their young before migrating to tropical regions for the winter season. The Neotropical song birds that are insectivorous have had an important role in the evolutionary development of the temperate forests and continue to have a critical role in controlling insect populations in these forests. A decline in the numbers of these birds is seen as a decline in the quality of the forest environments.

The rapidly developing pace of forest fragmentation in both areas—the tropical wintering areas and the temperate forests of North America—has had a significant impact on and resulted in a decline of the numbers of Neotropical migrants (see Figure 3.11). Forest fragmentation due to residential and infrastructure developments, logging, conversion to nonforest usages, and selective replanting of forests has interfered with the requirements of many Neotropical birds for large continuous woodlands in which to raise their young. Fragmentation of forests forces their nesting areas to the edges of forest fragments, exposing them to various predators. The consequence has been a slow decline in the numbers of many Neotropical birds.

Figure 3.11 Carolina Parakeet, an extinct Neotropical migrant. (© *by Susan L. Woodward, used by permission.*)

spring season to breed. Among them are the Wood Thrush and the Hermit Thrush, the Rose-breasted Grosbeak, and the Scarlet Tanager. A number of different hawks regularly migrate to breed in the northern half of the biome; in colder winters, those inhabiting the northern two-thirds of the biome migrate south. Hawks that migrate from the northern sections of the biome include the Red-tailed Hawk, the Broad-winged Hawk, and the Red-shouldered Hawk. Perhaps the majority of the birds found here in spring and summer season are migrants. A great number of birds, however, are year-round residents of the biome. Among them are such commonly observed species as the Black-capped Chickadee, the White-breasted Nuthatch, the Downy and Hairy woodpeckers, the Pileated Woodpecker, and the Blue Jay. Most owls are year-round residents as well, although they will move somewhat according to the availability of their prey. Among them are the Great-horned Owl, the

Barred Owl, and the Screech Owl. Ground-dwelling large birds are common in some areas, whereas in others they seem to be rare. These include the Turkey and the Ruffed Grouse.

Among the mammals of the biome the short-tailed shrew is with great likelihood the most abundant one throughout these forests. It is small and typically feeds within the forest litter. People usually overlook this mammal since it is not easily noticed. Quite visible, however, at dusk throughout the region are numerous bats. Common among these are the eastern pipistrelle and the big brown bat. Among the rodents of this biome, the most visible ones are the gray squirrel, and the somewhat less often seen fox squirrel. The eastern chipmunk is nearly as common but seems to live in denser underbrush than do squirrels. Among the mice are the white-footed mouse and the deer mouse. Both are relatively abundant throughout the biome. Common omnivores include the striped skunk and the raccoon. The black bear inhabits these forests but is not as frequently encountered as skunks or raccoons. Among the small carnivores are long-tailed weasel and red fox. Not quite as abundant as these two is the bobcat. The white-tailed deer, an abundant large herbivore, prefers the forest edge and young forest stands. It has adapted well to the human-dominated habitats of this forest region; indeed, it has become a pest in the eyes of many suburban gardeners. Less common and only recently reintroduced in the central portions of the biome (West Virginia) is the wapiti or elk. Adaptations to the changing habitats of this region have been more difficult for other species. The cougar or mountain lion was deliberately hunted and nearly extirpated throughout this biome. Sightings of cougars ranging from Georgia to Wisconsin have been reported. As of 2007, however, scientific proof of its successful comeback was lacking. The wolf also was deliberately extirpated over much of the area. Attempts to reintroduce wolves to Smoky Mountains National Park are in progress, but there is no widespread distribution of wolves in the central and southern portions of this biome as of this date. Perhaps the rapid expansion of the populations of the smaller relative of the wolf, the coyote, is a recent adaptation by wildlife to changes in habitat. The coyote is a recent immigrant from the western prairies. An animal of more open habitats, it thrives in the patchwork of forests, open agricultural lands, and suburban regions that characterize much of the eastern United States today.

East Asia

The Temperate Broadleaf Deciduous Forest Biome in East Asia is depicted in Figure 3.12. It once covered the East Asian portion of the continent as an unbroken forest between the boreal forest to the north and the tropical rainforest to the south. The geology of this broad region is one that has been relatively stable since the Paleozoic Era (570–245 mya) and few surface changes have occurred since, except by normal processes of weathering and fluvial erosion. In contrast to the situation

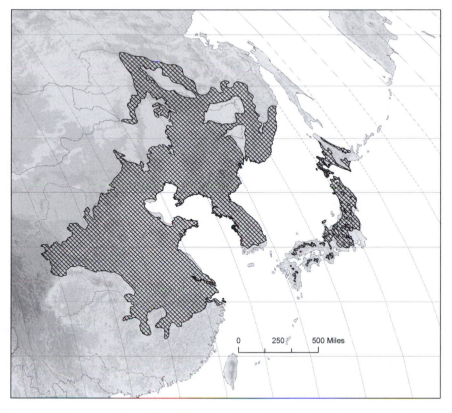

Figure 3.12 Temperate Broadleaf Deciduous Forest Biome in East Asia. *(Map by Bernd Kuennecke.)*

in the western and northern parts of the Eurasian Continent and North America, the Pleistocene Epoch (1.6 million to 10,000 years ago) left the geographic region of this biome largely untouched by glaciation. One result of this long-term stability is that several older lineages of plants have been preserved here and only here. Relict genera (they have survived from former climatic conditions) from the Tertiary Period (66–1.6 mya) are still found in this region of the biome, among them two ancient deciduous gymnosperms, the ginko and the dawn redwood.

The environmental complexity of this part of the biome is related to the fact that the geologic surface has not been severely disturbed by glaciation or recent tectonism and that several different types of physiographic features occur within this region, such as mountain ranges, coastal plains, and alluvial floodplains. These factors have contributed, at least in part, to the high diversity of plant communities in this part of the biome. The modern temperate broadleaf deciduous forests in East Asia today also reflect human impacts. The natural forest has been destroyed over most of its former natural extent by millennia of human activities, such as intensive agriculture and harvesting of wood for fuel. Despite this, some significantly large tracts of the natural forest stands still exist in remote mountain areas, as well as in many old reserves that were established in the past around important temples. These remnants of the former forests allow at least a partial reconstruction of the original forest associations of the region.

Three forest associations are identified in China. The first is located in the southern Yangtze Valley. The climate here is characterized by summers with very hot temperatures and high humidity levels brought about by the southwest monsoons. The winter periods are typically cold and dry, a result of the high-pressure currents of cold air that flow into the region during the winter months. This is the richest part of the biome in terms of its biological diversity, the mixed mesophytic forest. Throughout this subregion of the biome, oaks are dominant and include sawtooth oak, oriental white oak, Chinese white oak, and Chinese cork oak, among others. They are joined by chestnut, sweet gum, hornbeam, basswood, sassafras, and one of the relict broadleaf trees, a member of an ancient genus of walnuts. Two other relict trees, mentioned above, the dawn redwood and the ginko, both continue to exist throughout this forest region. The mixed mesophytic forest association is bordered to the south by a subtropical broadleaf evergreen forest. To the north of the mixed mesophytic forest region between the latitudes of 32° 30′ N and 42° 30′ N, lies another warm temperate broadleaf deciduous forest, in which oaks also dominate. The species composition is somewhat different than that farther south in the mesophytic forests. Here among the dominant oaks are Daimyo oak, Liaotung oak, and Mongolian oak.

The most extensive forests of the Temperate Broadleaf Deciduous Forest Biome in East Asia are found in the northern and northeastern provinces of China. More than 20 genera of forest trees dominate in this warm temperate mixed northern hardwood forest. This diversity of species is exemplified by the occurrence of common trees that codominate in the canopy layer, trees such as oaks, maples, hornbeam, alder, walnut, poplar, hackberry, ash, and basswoods. The structure of this vegetation typically consists of several layers. A canopy layer contains the dominants. Beneath this uppermost canopy is a secondary layer of somewhat smaller trees, such as smaller maples, hornbeams, and mountain ash. The shrub layer beneath the sapling layer contains woody plants such as dogwood, euonymus, and spicebush. A rich herb layer is typically encountered under the shrub layer. This forest type extends from northern China into the Korean Peninsula. It is also the dominant forest association on the Pacific side of Kyushu, Japan, between 37° 30′ and 38° N.

The limits of the broadleaf deciduous forest in China are met at higher latitudes and elevations. In these northern and mountainous areas is a cool temperate deciduous broadleaf forest dominated by birches (*Betula* spp.). Northwestern Japan has a different type of cool temperate forest. It is a forest that is dominated by a canopy of beeches and large oaks. Beneath is a thick undergrowth of bamboo.

The tremendous impacts on the forests by several thousand years of cultural land use in this region have not only decimated and altered the natural forests to the extent that only a few remnants survive in sanctuaries, but also have had a thorough influence on the fauna of this biome. With the exception of wildlife that has persevered in remote regions, inaccessible slopes, and parks, or that has been protected for cultural reasons, few indications of what the natural diversity of the fauna once entailed remain. It is assumed that animal life was once similar to that of the western part of the Eurasian continent. Some well-known animals still occur, such as the giant panda, the Siberian tiger, the raccoon dog, the sable, the rhesus monkey, and deer, such as musk, sika, and red deer. Other mammals include the golden snub-nosed monkey, the Himalayan black bear, the leopard, the Chinese goral, and the wild pig. Several common birds occur, including three species of pheasant, the While-bellied Woodpecker, and the Fairy Pitta. Some birds are highly protected and in some cases quite rare, such as the Black Stork, the White-tailed Sea Eagle, the Great Bustard, the Mandarin Duck, and the Crested Sheld-duck. These animals, for the most part, have been protected for some time now, probably the reason for their survival to the present. The Red-crowned Crane and the White-naped Crane, both endangered species, take advantage of the Demilitarized Zone (DMZ) between North and South Korea and utilize this region as a critical stopover in their migration as well as a breeding ground.

Europe

The European sections of the biome are shown in Figure 3.13. The deciduous temperate forests of Europe were greatly affected by the Pleistocene Ice Ages and the movement of continental ice sheets across the northern portions of this part of

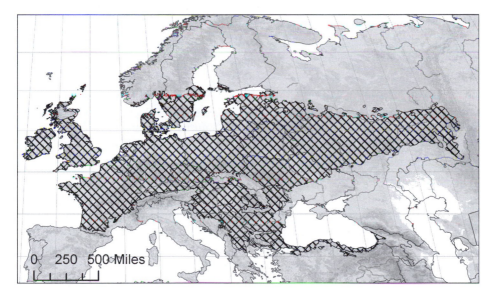

Figure 3.13 Temperate Broadleaf Deciduous Forest Biome in Europe. *(Map by Bernd Kuennecke.)*

Eurasia. The climatic influence of such large ice masses was as significant as the ice itself. The cold airmasses surrounding the continental ice sheets killed forests over extensive regions. A number of species that had adapted to these geographic regions prior to the Pleistocene Ice Ages suffered extinction because the east-west extent of mountains (the Alps) prevented migration southward. As a result, the European portions of the Temperate Broadleaf Deciduous Forest Biome today have the lowest diversity of plants of the four sections of the biome.

By far the most dominant tree over a wide range of geographic regions and correspondingly wide range of environmental conditions is the European beech. Due to human intervention by deliberate plantings, Scots pine is rapidly becoming the most common forest tree. A number of subregions are differentiated by those trees that accompany the beech within the forest associations, particularly the many oaks. This part of the Eurasian continent has a relatively high degree of environmental variation, the result of climate, terrain, distance from the ocean, and altitudinal situation. Forest associations are commonly distributed along environmental gradients. In the western sections, a maritime temperate climate (Marine West Coast or Cfb) with cool summers and mild winters extends far inland (see Figure 3.14). In the eastern sections, with increased distance from the Atlantic, continental climates (Dfa, Dfb) with warm to hot summers and cold winters assume predominance.

These eastern sections typically have significant annual temperature ranges. The latitudinal position of the respective stands also has a major role in climatic differentiation, so great north-south differences in forest composition are found in this part of the biome. Plant geographers and botanists in Europe typically distinguish between two major regional forest types, which are delineated according to latitude: the cold-temperate forest region of Middle Europe and the warm-temperate sub-Mediterranean forest region located north of the Mediterranean zone. Each of these forest regions has several subdivisions or so-called forest provinces that are distinguished from each other longitudinally.

The plant associations in the Middle European forest region have beech as their primary component, with pedunculate oak or sessile oak as an important part of the mix (see Figure 3.15). Throughout the cool, humid maritime climate regions of the Atlantic coast, the North Sea coast, and Great Britain, is the Atlantic province, where the pedunculate oak was once dominant, but has yielded to beech in those areas where forests have been left intact. The sub-Atlantic province extends away from the northern coastal areas of the continent and ranges to the Elbe River in the east and across the Massif Central of France to the Pyrenees Mountains on the border between France and Spain. In the sub-Atlantic province, sycamore, sessile oak, and lime join beech as canopy trees. The central European province, which spreads eastward from the Sub-Atlantic across Eastern France, the Benelux countries, and Germany to central Poland, finds beech and sessile oak at their current easternmost limits. From central Poland across the Ukraine to the southern Ural Mountains, the deciduous forest is dominated by the pedunculate oak. This is the sarmatic

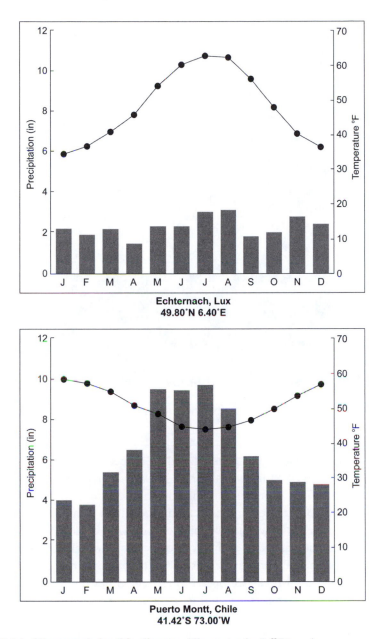

Figure 3.14 Climograph for Cfb climate. *(Illustration by Jeff Dixon.)*

province, which constitutes the fourth and last province of the Middle Europe forest region. This association lacks most of the trees found in either the Atlantic or the sub-Atlantic provinces. It is also a forest region that has suffered tremendous losses through conversion to agriculture and, in part, through modern forestry practices that favor only a few commercially valuable species (see Figure 3.16).

Figure 3.15 Beach-oak forests in Germany. *(© by L. Sue Perry, by permission.)*

The largest remaining section of these Middle European forests with largely natural plant associations is the Bialowieza Forest. It is a forest region that covers a mere 485 mi^2 (1,250 km^2) located astride the Poland-Ukraine border. Bialowieza is dominated by hornbeam. Other trees in this association include common ash, European alder, European aspen, Norway maple, birches, and wild apple. One oddity is the occurrence of some boreal plants, such as cloudberry and dwarf birch, as relics from the time when continental glaciers of Eurasia were retreating northward.

To the south of the Middle European forest region lies the sub-Mediterranean forest region. It is but a remnant of its former natural extent. This forest is dominated today mostly by oaks. In the past, other trees were of high importance in the canopy layer. Among them were trees such as Spanish chestnut, elms, and limes. Three provinces are distinguished within the sub-Mediterranean forest region. Western-most is the west sub-Mediterranean province, which is found across northern Spain, the Pyrenees (except for alpine association at high altitudes), and as a thin band of forests southwest of the Massif Central. The forests were once dominated by oak and boxwood. To the east is the middle sub-Mediterranean province, which once had a natural plant association distinguished by the absence of the dominant key species of the west province. Instead, eastern trees, such as hophornbeam, ash-elm, and the oak grew. Adjoining to the east is the east sub-Mediterranean province, formerly dominated by hornbeam, maple, and several oaks.

Animal life

The contemporary animal life of most of the forest regions of this biome in western, central, and Eastern Europe reflects the fact that most of the forests were decimated long ago to make room for agriculture and settlements. In addition, the forestry practices in one form or another, such as selectively leaving only commercially viable tree species, replanting with desirable species only, and "grooming" forested areas to provide better "hunting environment" for the nobility of the past (see Figure 3.16), have significantly altered the habitat for animals. Hunting in the past and at present favors some species, while others such as wolves, have been and still are deliberately eliminated. Most animals that have survived the decimation of European forests are animals that are as well identified as creatures of parks, hedgerows, and gardens as of the temperate broadleaf deciduous forests. Western and north-central Europe have a highly varied avifauna with several hundred species represented, including many that are rare and endangered. Many of the same families of birds encountered in American forests are characteristic in Europe. Birds include the nuthatch; tits such as the Great Tit and Blue Tit; woodpeckers such as the White-backed, Green, and Gray-headed; and thrushes such as Blackbird, Song Thrush, and the Nightingale.

Figure 3.16 Oak forest in planted rows, Reinhardswald, Germany. (© *by L. Sue Perry, by permission.*)

Nightingale

The nightingale received its name from the fact that it is often heard singing at night when other birds commonly are still. It is a bird slightly larger than most sparrows, and its song is relatively loud and full of trills. It was once a prized cage bird because of its melodious voice. One hears nightingales particularly during spring and summer nights in central Europe. It is the unpaired male nightingale that does most of the singing—it is not entirely clear whether the singing is done to attract a female or if it is a way to defend its territory. This is a bird that has adapted to life in cities and towns, where it sings even louder to overcome the noise of traffic and other human activity.

Forest flycatchers, such as the Red-breasted and Pied, seem familiar and exhibit a seemingly identical behavior to the flycatchers of North America, but they belong to an entirely different family of birds (Muscicapidae, versus Tyrannidae in the New World). When hiking in any of the forest regions across Europe during spring and summer, one typically catches the sound of one of the most often heard birds of the region, the Old World Cuckoo. For most listeners, the cuckoo's calling is rather more welcome than the sounds emitted by the bird's namesake—the cuckoo-clock.

Among the small mammals of the temperate deciduous forests of Europe those that are quite common, but perhaps not obvious, are the edible dormouse, the garden dormouse, the yellow-necked mouse, and the red-backed mouse or bank vole. Insectivores include the European shrew and the long-tailed shrew. Some characteristic mammals are perhaps well known, such as the Western European hedgehog, the ermine, and the European hare. Lesser midsize mammals include the Eurasian red squirrel, the Eurasian badger, and the European polecat. Larger mammals vary significantly in their spatial occurrence throughout this segment of the biome. The most common in the deciduous forests is clearly the roe deer. It is often observed at the edges of mature forest stands, in either young stands of trees or agricultural lands. Nearly as common in some of the hilly and mountainous forests are red deer and fallow deer.

Cuckoo

The Common Cuckoo (formerly referred to as European Cuckoo) is a member of the cuckoo order of birds. Its characteristic call (imitated by the cuckoo-clocks) can be heard in western Eurasian forests along their periphery throughout late spring and summer. The numerous folktales attached to the call of the cuckoo vary among cultural groups. In Germany, the first call of the cuckoo after winter is supposed to herald the arrival of spring. In eastern Europe, some folks believe that the number of calls of a cuckoo a person hears is indicative of the number of years that person has to live. In some regions, the call of the cuckoo is to announce to the listener that he or she will have a successful year.

The Common Cuckoo is a brood parasite. This means the female cuckoo lays her eggs into the nests of other birds. In this the female specializes on a particular type of bird and quickly lays an egg into that bird's nest when the opportunity arises, replacing one of the host's original eggs. The incubation period is typically shorter for the cuckoo egg than for the other eggs in the host nest. When the cuckoo chick has hatched, it grows quickly and ejects the other eggs or hatchlings, always clamoring for all the food its foster parents can bring.

Wild boars also inhabit wooded areas, but they are relatively shy. They are mostly nocturnal and not easily observed, except when disturbed at their daytime resting places. Wild boars have experienced a significant population explosion in the last two decades and are quickly becoming nuisance animals in the agricultural areas that are interspersed in the forests of Europe. Of course, boars inflict a significant amount of damage to young trees and roots on replanted forest stands if they can gain access (see Plate X). In the eastern provinces of this section of the biome, temperate broadleaf deciduous forests were home to the wisent and the wood tarpan well into the nineteenth century. In the twentieth century, the wisent was successfully reintroduced in protected areas in Eastern European forests. The tarpan, however, was a small horse that is now extinct in the wild.

South America

In South America, temperate broadleaf deciduous forests are restricted to a small region in southern South America (indicated on Figure 3.1). They are noteworthy, though, because of the uniqueness of the plant associations. The dominant climax trees are the deciduous southern beeches (*Nothofagus* spp.). This is a genus that is restricted to the lands of the former Gondwanan supercontinent.

In the south-central portions of Chile, southern beech forests occur in a region characterized by climatic conditions of significant winter precipitation and much less precipitation during the summer months (see Figure 3.14). (Note that seasons are reversed from the Northern Hemisphere.) In a region of mountain ranges that are exposed to the westerly winds from the Pacific, the Coast Ranges and the Andes cause orographic uplift. The windward sides of these mountains subsequently receive high annual precipitation, which increases with the altitude of the treeline in the Andes. In this region, the deciduous beech forests are found above the elevation of the mediterranean scrub vegetation that characterizes much of the area to the north and the lower elevations. The beech forest association is composed mostly of the roble. "Roble" is a Spanish word for oak, and the southern beeches at least superficially more strongly resemble oaks than northern beeches. Also found in this association are other

The Gondwanan Supercontinent

The term "Gondwanan" is commonly used to denote the distribution of living organisms that now occur in two discontinuous regions that were connected as parts of the southern supercontinent Gondwana in the geologic past. "Relicts" in the ecological literature are the still-existing remnants of an ecosystem that during the geologic past ranged over a large region but that today is restricted to relatively small areas. The genus *Nothofagus*, also known as the Southern Beeches, once ranged over the large regions of Gondwana, including what is today South America, Antarctica, New Zealand, and Australia. The *Nothofagus* forests that occur today in a small area of Chile (and another small area of New Zealand) are relicts of this past (see Plate XI). Another relict is *Araucaria*, a genus of evergreen coniferous trees, which, similar to *Nothofagus*, only occurs in the former Gondwanan region. More than 20 of the woody angiosperms of the temperate forests of South America have Gondwanan heritages.

southern beeches, such as hualo. The trees in these stands can reach average heights that vary from 65 to 80 ft (20 to 25 m). The association includes, in most cases, evergreen trees closely related to the mediterranean vegetation of Chile. Among them are alerce in the southern portions of the biome and the monkey-puzzle tree (north of 40° S) and Chilean cedar (north of 44° S) in the northern sections. The evergreen Chilean laurels of southern Chile are common throughout the region, along with lingue, olivillo, and avellano. Beneath the canopy of these forests is a sparse shrub layer often containing bamboo. In the northern part of the Coastal Range (33°–34° S) are small areas with stands of this forest type, sustained primarily by the moisture that precipitates from coastal fogs. Farther south, into central and south-central Chile, are roble-hualo forests, closely tied to the cooler and more humid south-facing slopes and ravines in the Coastal Range. They typically occur at elevations from 2,000 to 8,000 ft (650 to 2,500 m) in the Andes Mountains.

In the Coastal Range between 37° and 40° S in southern Chile and in southwestern Argentina, the deciduous *Nothofagus* forests and mixed *Nothofagus* forests occur. In the Central Valley, this forest type lies between 38° and 41° 30′ S. In the Andes, it is found between 36° and 40° 30′ S. Beyond the mediterranean conditions of central Chile, these areas experience no summer drought, and the trees have adapted to conditions that provide abundant year-round precipitation. Precipitation increases with latitude. Orographic precipitation occurs on the windward (western) side of the coastal mountains and in the Andes, because the uplift of westerly flowing airmasses off the Pacific causes a loss in the air's capacity to hold moisture. As a result, the windward slopes receive between 120 and 200 in (3,000 and 5,000 mm) of precipitation a year (see Figure 3.17).

The temperatures of these regions are strongly moderated by the maritime influence and vary by altitude. On lowland forest sites with well-drained soils, roble is dominant. In such areas it can attain heights of more than 130 ft (40 m) and diameters that exceed 6 ft (2 m). The tree possesses a reddish-colored wood that has proven to be extremely resistant to rot and decay, similar to the redwoods of northern California and the American chestnut. The decay-resistant properties of its wood have made this tree a sought-after raw material for fence posts, outdoor paneling, decks, exposed beams, and so forth. Roble is similar to the American chestnut in another way: its ability to resprout after it has been cut. Roble often grows with another deciduous southern beech, the rauli, also a commercially important tree. An evergreen southern beech is commonly found in association with roble and rauli. The understory beneath the tall southern beeches in this section of the biome includes typically smaller evergreen trees, such as laurel, ulmo, olivillo or palo muerto, avellano or Chilean hazelnut, and winter's bark, the winter-blooming holy tree of the native Araucanian people who lived in and around these forest regions before the arrival of Europeans. The shrub layer is commonly dominated by bamboos. At the highest elevations of forest, the southern beeches first mix with other Gondwanan trees, such as the evergreen conifer monkey-puzzle tree, and

Figure 3.17 *Nothofagus* beech rainforest, southern Chile. *(Photo courtesy of Antonio Carlos de Barros Correa, Federal University of Pernambuco, Brazil.)*

then reach their elevation-induced temperature limit and give way completely to these ancient evergreen conifers.

On Tierra del Fuego, cool temperate *Nothofagus* forests occur from 37° 30′ S to about 55° S. At the lower elevations of these mountainous islands, an evergreen southern beech dominates. At higher elevations, the deciduous lenga and ñirre dominate forest stands. The latter is a relatively small tree at mid-elevation levels, where it reaches heights of 30–50 ft (10–15 m). Near treeline ñirre forms a krumm-holz with heights of only 6–10 ft (2–3 m).

Human Impacts

The effect of human activity has been felt in all regions where expressions of the Temperate Broadleaf Deciduous Forest Biome occur. The historical timing and sequences of such impacts have been different, but all forests of this biome have been altered by becoming centers of agricultural development in their respective geographic regions. In the course of the past several thousand years, the temperate deciduous forest regions of East Asia, as well as Europe, have experienced widespread clearing of the forests to make room for agricultural fields for food production. In eastern North America, this process of clearing forests for agriculture began some 400 years ago. In South America, the clearing of broadleaf deciduous forest stands for conversion to agricultural lands is ongoing. While the outright eradication of forest stands to make room for other land use is quite dramatic and sudden, the use of these forests for extensive grazing, forage for domesticated animals (such as hogs), exploitation of trees for construction, charcoal and potash production, shipbuilding, leather tanning, and so on has been ongoing for thousand of years. In addition, the impact of using some trees while leaving others has had a tremendous effect—that is, the selective harvest of the best of the desirable trees has left other, less desirable ones for reseeding of the stands (negative breeding). The highly accessible lowland forests that once existed in China and Europe essentially have been converted to other land uses. In East Asia, only remnant forests are preserved in the less accessible mountains of China as well as in Japan and Korea. In the latter two, some significantly sized remnants of these forests survive. In North America, almost all of the very large areas of temperate broadleaf deciduous forests are second growth forests. Many have become established on former farm fields that had been left fallow, often because of excessive soil erosion, steepness of slopes, or marginal crop production. Only few area of the former natural forests that were the least suited for cultivation and grazing have some remnants of old-growth forests today. These areas were either too steep, too poor and thin in their soils, or too wet to attract agricultural use.

The alterations of forest regions in Europe began at least 5,000 years ago with the clearing of forest stands through burning and girdling to convert them to farmland. At the beginning of the Bronze Age, several regions across Europe saw heathlands and blanket bogs beginning to replace forests as a result of frequent burning and sheep grazing. By that time, extensive, permanent agricultural settlements had developed across much of Europe, especially in the temperate and warm climate areas. Agricultural development in terms of techniques and crop yields supported growing human populations prior to the emergence of the Roman Empire. The Roman Period witnessed a tremendous expansion of settlements and thus increased clearing of forest areas. As population continued to grow, demand for agricultural products pushed the clearing of forests for conversion to agricultural land forward. Significant losses of populations due to the Black Plague resulted in the abandonment of portions of agricultural lands from the thirteenth through fifteenth centuries, particularly in central Europe. Further reductions in populations were experienced during the Thirty Years' War (1618–1648) along with abandonment of farmland and regeneration of forests on many of the formerly cleared lands. Extensive use of forest lands by agriculturalists as pasture for cattle, sheep, goats, and horses continued around the remaining settlement areas. This practice of use of forest for grazing certainly changed the composition and structure of the understory of the forests. Additionally, people intentionally protected those trees deemed beneficial and often propagated their

Plate I. Moose. *(Courtesy of Shutterstock #5163748. © Arnold John Labrentz.)*

Plate II. Wapiti. *(© Susan L. Woodward, used by permission.)*

Plate III. Caribou herd. *(Courtesy of Shutterstock #3445970. © Louise Cukrov.)*

Plate IV. Musk ox. *(© L. Sue Perry, by permission.)*

Plate V. Crossbill. *(Courtesy of Shutterstock #1763195. © Jason Vandehey.)*

Plate VI. Sequoia needles. *(© Susan L. Woodward, used by permission.)*

Plate VII. Avalanche scar in lodgepole pine stand, Rocky Mountains. (© *Susan L. Woodward, used by permission.*)

Plate VIII. Clear-cut Douglas fir forest in Oregon. (© *Susan L. Woodward, used by permission.*)

Plate IX. Ecotone between boreal and deciduous temperate forests. (© *L. Sue Perry, by permission.*)

Plate X. Wild boar in Germany—damage caused by rooting. (© *L. Sue Perry, by permission.*)

Plate XI. *Nothofagus* forest, Chile. *(Photo courtesy of Antonio Carlos de Barros Correa, Federal University of Pernambuco, Brazil.)*

Plate XII. Chamise in bloom, California. *(© Susan L. Woodward, used by permission.)*

Plate XIII. Chamise resprout after a fire, California. (© *Susan L. Woodward, used by permission.*)

Plate XIV. Cape reeds, South Africa. (© *Susan L. Woodward, used by permission.*)

Plate XV. Fire erica, Cape Province, South Africa. *(© Susan L. Woodward, used by permission.)*

Plate XVI. Protea, Cape Province, South Africa. *(© Susan L. Woodward, used by permission.)*

Figure 3.18 Oak sprouts after coppicing. *(© Bernd Kuennecke.)*

dominance. Throughout much of Europe during that time, oak and beech trees were protected because people often survived the winters on hogs that had been fattened on acorns and beech nuts. Oak timbers were preferred for housing construction. Oak was the preferred material for fence posts and was the wood of choice for firewood and for making charcoal. The bark of oak trees was used in the process of tanning leather hides. Until recent times, peasant farmers in much of Europe planted oaks along driveways and country lanes as well as fences to guarantee a ready supply. Farmers typically utilized the leaves of some broadleaf trees as supplemental forage for their livestock. As a result, trees were commonly coppiced through the practice of harvesting whole branches that were then fed to livestock. This practice inadvertently favored some trees and caused the elimination of others. Those trees that will resprout from their roots or stumps, such as oak, lime, ash, hornbeam, and hazel, will survive such harvesting practices for forage (for example, coppicing) (see Figure 3.18).

Others, such as beech, do not resprout from roots or stumps, and this practice eliminated them from the landscape over large areas. Another tremendous impact on those forest areas that were not converted often resulted from the land tenure. The nobility of the state and the church typically controlled the access to the forest lands they owned. Peasants were charged user fees for grazing, gathering firewood, occasional timber harvesting, and, in some cases, hunting of small game. Large game hunting was reserved for the landowners. The landowner also determined the trees to be favored through selective harvest and any reforestation activities (see

Figure 3.16). Such determination was typically driven by the perceived habitat requirements for the big game the landowner wanted to hunt. Limiting access to the remaining forests in the Middle Ages caused the open lands or commons to come under severe pressure from overuse. Trees were mostly eliminated from such commons, converting them into barren, treeless areas. The impact on forests throughout Europe in the past several thousand years has been so severe that the only relatively unaltered forest region left in this expression of the biome is the Bialowieza National Park in eastern Poland. Another forest area in Europe that might be classified as largely natural in its forest composition is the Hainichen National Park in east-central Germany.

Reforestation in Europe has a history that goes back to the fourteenth and fifteenth centuries, initially as efforts to restore hunting areas for the nobility. Replanting logged-over and abandoned agricultural land with commercially desirable trees began in earnest in several areas of Europe toward the end of the eighteenth century. Reforestation was undertaken not just to grow timber and provide forested hunting areas, but also to produce other wood products (see Figure 3.16). In Germany, for example, oaks were planted when the Napoleonic wars prevented tanning bark to be shipped from The Netherlands. The end of the eighteenth and beginning of the nineteenth centuries witnessed the desirability of the reestablishment of wooded areas across much of Europe. Agricultural practices were, in part, responsible for a decrease in use of forest trees as forage—for example, the white potato had been introduced from the Andes of South America and became widely used as a winter food for livestock. This, of course, ended the need to harvest tree branches and young trees for livestock feed. Simultaneously, the previously existing need for large forest areas to graze sheep declined rapidly when wool production shifted from central and eastern Europe to Ireland and Australia. This was also a time of agricultural innovations and shifts in agricultural practices. The introduction of the three-field system, new crop varieties, and the growing practice of applying fertilizers to agricultural fields all made possible an increase in crop yields per acre, reducing the need for agricultural land, even when populations were expanding in Europe. Agricultural fields with marginal crop yield were abandoned and either left to be reforested by natural succession of plants or planted with tree seedlings of desirable species. Figure 3.19 shows an old beech forest in the Harz Mountains of Germany.

Prior to the Industrial Revolution one major use of forests was to produce charcoal. The industrial shift to coal resulted in a sharp decrease for the need of charcoal as a major fuel source. Industrial manufacturing favored stone (brick) and steel over timber as a construction material. All of this resulted in a decreasing need for the use of the forests. Reforestation was regarded as an economical use of open land. Many of the trees that were planted in the early nineteenth century were not the broadleaf deciduous ones, but the faster-growing spruce, pine, and larch that were desirable for rapid construction of buildings, particularly the supporting structures for roofs.

Figure 3.19 Old beech forest, originally replanted, Harz Mountains, Germany. (© *by L. Sue Perry, by permission.*)

The situation in eastern North America in terms of human impacts has been ongoing for some time as well, although we have a tendency to see this as something that has happened only since the arrival of European settlers. More than 50 years ago, an authority on these forests speculated that it was difficult to discern the original forest cover, since human activities had changed the forest composition significantly. The temperate broadleaf deciduous forests of eastern North America were occupied by a native population as much as 3,000 years ago. The archaeological record of these peoples suggests that they engaged in hunting and gathering the varied foods of the forests in addition to farming crops. They used fire rather commonly as a technique to clear the forests for temporary field plots for the cultivation of maize, squash, and beans. In addition, American Indians used fire as a means of driving game toward a group of hunters. Fire was also a tool used to clear brush and to control vegetation around their settlements, possibly creating defensive clearings to prevent any type of ambush by tribes not friendly to them. The evidence that Indians used fire rather purposefully for vegetation control is recorded in reports by early English settlers 400 years ago. These accounts state that the forests were free enough of shrubs and saplings to allow room for a stagecoach to be driven through the forests. The arrival of English and other European settlers heralded the beginning of plow agriculture in eastern North America. New crops were introduced, such as wheat, oats, and other Old World grains, as well as domestic livestock, among which hogs, sheep, and cattle were most significant in terms of affecting the landscapes. The new settlers quickly "Europeanized" the landscape through clearing of forests. By establishing settlements, they created a continuously shifting mosaic of cultivated fields, pastures, and woodlands (on the poorer soils and the steeper slopes). They cut timber for building materials, for defensive stockades, for the production of potash for soap-making and wool cleaning, to make charcoal for the smelting of iron and "cooking" of ores to win lead, and to strip trees of their bark for the tanning of leather. Hogs were driven into the woods to fatten up on chestnuts, acorns, and hickory nuts. Following the practices of Europe of the time, cattle and sheep were grazed in the woods, often in an open-range fashion. After several years of intensive use and no fertilization, cultivated land and pasture land often were left fallow and allowed to revert to a forest of sorts. Typically, such woodland areas would be cleared again for fields and pasture. The second half of the nineteenth century witnessed other assaults on the forests. The construction of railroads extended the access for lumbermen into previously remote locations and provided them with a means of transporting cut lumber from such areas to the markets in the settlements. The iron industry reached the Valley and Ridge physiographic province in the 1800s. The demand for charcoal grew tremendously. The coal mines that were established to extract coal as a more energy-efficient fuel for the iron industry led to a huge demand for mine timbers to support the shafts. For one purpose or another all but the most inaccessible forests and those that did not hold promise for agriculture throughout eastern North America were stripped of their trees at one time or another.

The composition of animal life in the forests of eastern North America also changed drastically with the European settlement of the region. The cultural values of the new settlers demanded that wolves and cougars be extirpated. Large herbivores like elk and bison were regarded as too much competition for the domesticated cattle, and they served as a handy source of meat. The population levels of the large herbivores were driven down rapidly, and they survived only in remote regions and the neighboring boreal forests to the north. The Carolina Parakeet and the Passenger Pigeon were both considered agricultural pests and a nuisance. Deliberate extermination of the former and overhunting of the latter rendered them extinct in the nineteenth century. Introductions of some species into the region, such as the accidental release of wild boars, have added to the mix of animals. At the same time, they have introduced a rather destructive species into this environment.

The forests of eastern North America appear today to be quite extensive. However, most of the stands are quite young (60–100 years). Additionally, the composition of these forests most certainly does not replicate the forests that existed in 1600. The modern forests are the result of numerous factors. Several periods of farm abandonment have occurred, associated with the opening of more productive lands in the western United States, with the Civil War, with the increasing mechanization of agricultural techniques, with excessive soil erosion on steep slopes, and with the Great Depression of the 1930s. The American chestnut once was the largest tree in the eastern forests and was overwhelmingly dominant in some stands.

Conversion of forests in mountainous areas has always carried the danger of destructive flooding after stripping the land of its the protective forest cover. Recognition that some form of forest restoration and subsequent management was necessary to reduce the incidences of excessive flooding due to eradication of forest cover in mountainous watersheds came early in the twentieth century. One manifestation of such new attitudes was the passage of the Weeks Act in 1911, which established much of the current system of national forests that we have today. Ten of the newly created national forests were located in the deciduous forests of the Appalachian Highlands of eastern North America.

The human impact on the forests of eastern North America continues, including those areas newly reforested either by natural regeneration or by planting of seedlings. The eastern forests are today under attack by air pollution, introduced insects, introduced wildlife, and various introduced diseases. Almost any forest stand within the temperate broadleaf deciduous forests of eastern North America exhibits visible damage in terms of crown and branch dieback. Premature leaf drop is common and growth has been slowed; in some stands, root decay is evident, and forest trees have increased rates of mortality. The die-offs began in the early 1960s and accelerated in the 1980s, when white oaks, beech, yellow buck-

Wild Boar

The wild boar (*Sus scrofa*) is native through the temperate, subtropical, and even some tropical regions of Eurasia, as well as across much of North America. It was introduced to the Americas, New Zealand, and Australia as an animal for hunting. Its relative, the domestic pig, was also introduced to the New World, where it was formerly maintained in free-roaming herds. Some of the domestic pigs escaped the control of people and became feral, and some feral stock interbred with introduced boars. Boars are not found in dry desert regions or in high alpine zones.

The wild boar is an omnivorous mammal with an appetite for almost anything: farm crops, nuts, berries, small animals, carrion, seedlings, acorns, and beechnuts all are suitable foods. Wild boar populations have been growing rapidly in the past two decades in central Europe in part because of reduced hunting pressures, but also because of the reforestation of formerly open fields, which provide them with expanded habitat. Wild boar are destructive (see Plate X) in newly planted forests because they dig up seedlings for food and destroy them by rooting for insects and grubs. They may prevent natural reforestation from nuts, acorns, and seeds by consuming so many of them. In farming areas, wild boars are particularly damaging to row crops (potatoes, beets, corn, and so on). Hunting them in such agricultural environments often pushes them back into the forests.

eye, tulip poplar, dogwood, sassafras, hickories, walnuts, white basswood, hemlock, and sugar maple all showed signs of decline throughout many forest stands. The forests of eastern North America are located largely within and downwind of the old manufacturing belt of the United States. Many of the old industries are still active today. Numerous new ones have been established within this belt throughout the past decades. Coal-fired power generation plants are located within the forests surrounding the megalopolis of the eastern seaboard, a densely urbanized and highly populated region that stretches along the Atlantic Coast from Virginia Beach, Virginia, to Portland, Maine. The emission of acid-forming gases (particularly sulfur dioxide) into the air by industries, vehicles, residential heating plants, and so forth results in acids when combined with moisture. As a result, acid deposition in rain, fog, rime ice, and snow can directly damage the leaves of trees. Other impacts of air pollution result from the addition of excess nitrogen and sulfur, which changes the chemistry of the soil, increases the acidity of the soils layers, and thus slows the bacterial decay processes vital to nutrient cycling within the root zone of trees. The increased acidity of soils also increases available aluminum to levels that may be quite harmful to forest plants. The exhaust from vehicles contributes to the development of high doses of ground-level ozone, which negatively affects trees. The sum total of all such stresses on trees has the effect of weakening them and renders them more susceptible to attacks by diseases and insects than what would naturally be the case. When mature trees of the canopy layer die, it allows for more sunlight to penetrate below the canopy layer. This, in turn, causes the forest floor to be warmed more, and it dries the materials of the forest litter. This makes it difficult for salamanders and frogs to exist, since both require the moisture in the forest litter. Dry conditions in the forest litter also create adverse conditions for the regeneration of the canopy trees from seeds. The resultant slowing of tree growth translates into less flowering and less fruiting. Subsequently, it affects the populations of those animals that depend on acorns and nuts. In the forests of eastern North America, susceptibility to diseases and insect attacks are becoming evident to the visitors of these forests. The sapstreak disease in sugar maples that is caused by the fungus *Ceratocystis coerulescens* is but one of these diseases. Hemlocks are defoliated and killed by the hemlock woolly adelgid (*Adelges tsugae*). The so-called beech bark disease occurs when the bark of beech trees is damaged by the beech scale (*Cryptococcus fagisuga*) and when the fungi (*Nectria coccinea* or *N. galligena*) invade through such openings. Anthracnose (*Discula destructiva*) is a destructive fungus that is killing dogwoods in eastern forests. Throughout the past decades, the caterpillar of the introduced gypsy moth (*Lymantria dispar*) has been active in the defoliation of oaks, hickories, and other trees, and it contributes substantially to the weakening and dying of trees in the areas it has infested. The gypsy moth has been expanding its territory rather steadily, particularly along transportation routes, hence the name "gypsy." Other significant impacts that threaten the temperate deciduous forests in eastern North America are the direct impact of urban sprawl, residential development in forest regions, the development of transportation routes through forests regions, the building of pipelines and electric transmission lines through forests, and the continued clear-cutting of forest stands to produce lumber and pulpwood. In addition, we are realizing to an increasing degree the ongoing climate change caused by global warming tendencies. Warming will cause a renewed northward migration of the beech into the southern areas of those forests currently classified as boreal.

The temperate deciduous *Nothofagus* forests of South America have been affected by human impacts only for the past 100 years. The initial impacts were primarily related to clearing of forests to convert the land to agricultural usages. However, the lumber value of roble and rauli were also realized, and both were cut for use of the timbers in shipbuilding as well as for the production of charcoal. Although of relatively short duration, the impacts have been significant in this expression of the biome. Many of the original stands of the temperate deciduous *Nothofagus* forests of South America are gone. Reforestation, where attempted, has typically involved the replanting of stand areas with the nonnative Monterey pine. The last few years have witnessed a growing interest in utilizing native trees in the reforestation attempts in the region, and the production of commercially important native trees by tree nurseries to replant these forests is increasing.

Further Readings

Braun, E. Lucy. 1950. *Deciduous Forests of Eastern North America*. Philadelphia: Blakiston.

FAO. 2001. Global Ecological Zoning for the Global Forest Resources Assessment. http://www.fao.org/docrep/006/ad652e/ad652e00.htm.

Appendix

Biota of the Temperate Broadleaf Deciduous Forest Biome (arranged geographically)

North America

Oak-chestnut Forest Type

Trees

American chestnut[a]	*Castanea dentata*
Chestnut oak	*Quercus prinus*
White oak	*Quercus alba*
Tulip poplar	*Liriodendron tulipifera*
White basswood	*Tilia heterophylla*
American beech	*Fagus grandifolia*
Yellow birch	*Betula alleghaniensis*
Sugar maple	*Acer saccharum*
Mountain ash	*Sorbus americana*
Yellow buckeye	*Aeschulus octandra*
Eastern hemlock	*Tsuga canadensis*
Red spruce	*Picea rubens*

Understory trees and shrubs

Maple-leaved viburnum	*Viburnum alnifolium*
Mountain maple	*Acer spicatum*
Striped maple	*Acer pensylvanicum*
Redbud	*Cercis canadensis*
Flowering dogwood	*Cornus florida*
Shadbush or Allegheny serviceberry	*Amelanchier laevis*
Silverbells	*Halesia carolina*
Spicebush	*Lindera benzoin*

Vines and climbers

Poison ivy	*Toxicodendron radicans*
Wild grape	*Vitis* spp.

124

Forbs

Hepatica	*Hepatica americana*
Mayapple	*Podophyllum peltatum*
Skunk cabbage	*Symplocarpus foetidus*
Spring beauty	*Claytonia virginica*

Note: [a]Extinct as a canopy tree.

Oak-Hickory Pine Forest

Trees

White oak	*Quercus alba*
Post oak	*Quercus stellata*
Black oak	*Quercus velutina*
Scarlet oak	*Quercus coccinea*
Chestnut oak	*Quercus montana*
Chinkapin oak	*Quercus muhlenbergii*
Northern pin oak	*Quercus ellipsoidalis*
Red oak	*Quercus borealis*
Northern red oak	*Quercus rubra*
Post oak	*Quercus stellata*
Blackjack oak	*Quercus marilandica*
Willow oak	*Quescus phellos*
Water oak	*Quescus nigra*
Overcup oak	*Quercus lyrata*
Shagbark hickory	*Carya ovata*
Bitternut hickory	*Carya cordiformis*
Red hickory	*Carya ovalis*
Shellbark hickory	*Carya laciniosa*
Scrub hickory	*Carya floridana*
Sugarberry	*Celtis laevigata*
Black walnut	*Junglans nigra*
White basswood	*Tilia heterophylla*
Sassafras	*Sassafras albidum*
Sweetgum	*Liquidambar styraciflua*
Shortleaf pine	*Pinus echinata*
Virginia pine	*Pinus virginiana*
Loblolly pine	*Pinus taeda*
Longleaf pine	*Pinus palustris*
Eastern hemlock	*Tsuga canadensis*

Understory trees and shrubs

Red maple	*Acer rubrum*
Sourwood	*Oxydendron arboretum*
Black tupelo	*Nyssa sylvatica*
Flowering dogwood	*Cornus florida*

(*Continued*)

Lowbush blueberry	*Vaccinium angustifolium*
Sparkleberry	*Vaccinium arboreum*
Rusty blackhaw	*Viburnum rufidulum*
Blackhaw	*Viburnum prunifolium*
Maple-leaf viburnum	*Viburnum acerifolium*
Sheep-laurel	*Kalmia angustifolia*
Mountain laurel	*Kalmia latifolia*
Blue huckleberry	*Gaylussacia frondosa*
Black huckleberry	*Gaylussacia baccata*

New Jersey Pine Barrens

Trees

Pitch pine	*Pinus rigida*
Shortleaf pine	*Pinus echinata*
Atlantic white cedar	*Chamaecyprus thyoides*
Blackjack oak	*Quercus marilandica*
Bear oak	*Quercus ilicifolia*
Dwarf chinkapin oak	*Quercus prinoides*
Post oak	*Quercus stellata*
Chestnut oak	*Quercus montana*

Understory trees and shrubs

Black tupelo	*Nyssa sylvatica*
Blue huckleberry	*Gaylussacia frondosa*
Black huckleberry	*Gaylussacia baccata*
Mountain laurel	*Kalmia latifolia*
Sheep laurel	*Kalmia angustifolia*
Red maple	*Acer rubrum*
Cranberry	*Vaccinium oxycoccus*
Bear oak	*Quercus ilicifolia*
Dwarf chinkapin oak	*Quercus prinoides*
Box sandmyrtle	*Leiophyllum buxifolium*

Mixed Mesophytic Forests

Trees

American chestnut[a]	*Castanea dentata*
Northern red oak	*Quercus rubra*
White basswood	*Tilia heterophylla*
American hornbeam	*Carpinus caroliniana*
Cucumbertree	*Magnolia acuminata*
Sugar maple	*Acer saccharum*
Tulip poplar	*Liriodendron tulipifera*
Black walnut	*Juglans nigra*
White ash	*Fraxinus americana*

Understory trees and shrubs

Bigleaf magnolia	*Magnolia macrophylla*
Fraser magnolia	*Magnolia fraseri*
Umbrella magnolia	*Magnolia tripetala*
Black tupelo	*Nyssa sylvatica*
Flowering dogwood	*Cornus florida*
Pawpaw	*Asimina triloba*
Redbud	*Ceercis canadensis*
Sourwood	*Oxydendron arboretum*
Striped maple	*Acer pensylanicum*

Note: [a]Extinct as a canopy tree.

Western Mesophytic Forests

Trees

American beech	*Fagus grandifolia*
Sugar maple	*Acer saccharum*
Blue ash	*Fraxinus quadrangulata*
White ash	*Frasinus americana*
White oak	*Quercus alba*
Bur oak	*Quercus macrocarpa*
Northern red oak	*Quercus rubra*
Black oak	*Quercus vetulina*
Basswood	*Tilia heterophylla*
Shagbark hickory	*Carya ovata*
Bitternut hickory	*Carya cordiformis*
Red hickory	*Carya ovalis*
Shellbark hickory	*Carya laciniosa*

Cedar Glades

Trees

Red cedar	*Juniperus virginiana*
Black oak	*Quercus vetulina*
Chinkapin oak	*Quercus muehlenbergii*
Post oak	*Quercus stellata*
Northern red oak	*Quercus rubra*
Shagbark hickory	*Carya ovata*
Winged elm	*Ulmus alata*

Northern Hardwood Forests

Trees

American beech	*Fagus grandifolia*
Sugar maple	*Acer saccharum*

(*Continued*)

Northern red oak	*Quercus rubra*
Paper birch	*Petula papyrefera*
Quaking aspen	*Populus tremuloides*
Bigtooth aspen	*Populus grandidentata*
Balsam fir	*Abies balsamea*
Eastern hemlock	*Tsuga canadensis*
White pine	*Pinus strobes*
Red spruce	*Picea rubens*

Southeast Evergreen Forests

Trees

Longleaf pine	*Pinus palustris*
Loblolly pine	*Pinus taeda*
Slash pine	*Pinus elliottii*
Blackjack oak	*Quercus marilandica*
Bald cypress	*Taxodium distichum*
Cherrybark oak	*Quercus falcata*
Laurel oak	*Quercus laurifolia*
Live oak	*Quercus virginiana*
Overcup oak	*Quercus lyrata*
Post oak	*Quercus stellata*
Southern red oak	*Quercus falcate*
Turkey oak	*Quercus laevis*
Water oak	*Quescus nigra*
Black walnut	*Juglans nigra*
Persimmon	*Diospyros virginiana*
Sweetgum	*Liquidambar styraciflua*
Southern magnolia	*Magnolia grandiflora*
Swamp chestnut oak	*Quesrcus michauxii*
Sweetbay magnolia	*Magnolia virginiana*
Sycamore	*Platanus accidentalis*
Water hickory	*Carya aquatica*
Water elm	*Planera aquatica*
Willows	*Salix* spp.

Understory trees and shrubs

American holly	*Ilex opaca*
Ogeechee tupelo	*Nyssa ogeche*
Water tupelo	*Nyssa aquatica*

Animals of the Temperate Broadleaf Forests of North America

Mammals

| Short-tailed shrew | *Blarina brevicauda* |
| Big brown bat | *Eptesicus fuscus* |

Eastern pipistrelle	*Pipistrellus subflavus*
Gray squirrel	*Sciurus caroliniendis*
Fox squirrel	*Sciurus niger*
Eastern chipmunk	*Tamias striatus*
White-footed mouse	*Peromysucs leucopus*
Deer mouse	*Peromyscus manuclatus*
Woodchuck or Groundhog	*Marmota monax*
Red fox	*Vulpes vulpes*
Wolf	*Canis lupus*
Coyote	*Canis latrans*
Black bear	*Ursus americanus*
Raccoon	*Procyon lotor*
Long-tailed weasel	*Mustela frenata*
Striped skunk	*Mephitis mephitis*
Bobcat	*Felis rufus*
Wapiti or Elk	*Cervus elaphus*
White-tailed deer	*Odocoileus virginianus*
Wood bison	*Bison bison*

Birds

Red-tailed Hawk	*Buteo jamaicensis*
Red-shouldered Hawk	*Buteo lineatus*
Broad-winged Hawk	*Buteo platypterus*
Barred Owl	*Strix varia*
Great-horned Owl	*Bubo virginianus*
Screech Owl	*Otus asia*
Turkey	*Meleagris gallopava*
Ruffed Grouse	*Bonasa umbellus*
Downy Woodpecker	*Dendrocopus pubescens*
Hairy Woodpecker	*Dendrocopus villosus*
Pileated Woodpecker	*Dryocopus pileatus*
Blue Jay	*Cyanocitta cristata*
Red-eyed Vireo	*Vireo olivaceous*
White-breasted Nuthatch	*Sitta carolinensis*
Black-capped Chickadee	*Poecile atricapillus*
Hermit Thrush	*Hylocichla guttata*
Wood Thrush	*Hylocichla mustelina*
Black-and-white Warbler	*Mniotilta varia*
Chestnut-sided Warbler	*Dendroica pensylvanica*
Yellow-throated Warbler	*Dendroica dominica*

Reptiles

Common garter snake	*Thamnophis sirtalis*
Common kingsnake	*Lampropeltis getula*

(*Continued*)

Racer   *Coluber constrictor*
Rat snake   *Elaphe obsolete*
Timber rattlesnake   *Crotalus horridus*
Copperhead   *Agkistrodon contortix*
Eastern fence lizard   *Sceloporus undulatas*

Amphibians
Peaks of Otter salamander   *Plethodon hubrichti*
Red-backed salamander   *Plethodon cinereus*

East Asia

Mixed Mesophytic Forest

Southern trees
Dawn redwood   *Metasequoia glyptostroboides*
Ginko   *Ginko biloba*
Sawtooth oak   *Quercus acutissima*
Oriental white oak   *Quercus aliena*
Nakai   *Quercus chenii*
Chinese white oak   *Quercus fabri*
Chinese cork oak or   *Quercus variabilis*
  Oriental oak
Basswood   *Tilia henryana*
Chinese chestnut   *Castanea seguinii*
Hornbeam   *Carpinus fargesii*
Sweet gum   *Liquidamabar formosana*
Relict member of walnut   *Platycarya strobilacea*
  family
Chinese sassafras   *Sassafras tzumu*
Pistachio   *Pistacia chinensis*
Red pine   *Pinus densiflora*

Northern trees
Daimyo oak   *Quercus dentata*
Liaotung oak   *Quercus liaotungensis*
Mongolian oak   *Quercus mongolica*
Oak (bao li)   *Quercus serrata*
Lime (linden)   *Tilia* spp.
Korean hornbeam   *Carpinus tschonoskii*
Japanese hornbeam   *Carpinus laxiflora*
Japanese maple   *Acer palmatum*
Shantung maple   *Acer mono*
Korean ash   *Fraxinus rhynchophylla*
Walnut   *Juglans mandshurica*
Cypress   *Paltycladus orientalis*

Korean white cedar	*Thuja koraiensis*
Korea fir	*Abies koreana*
Manchurian fir	*Abies holophylla*
Armandi pine	*Pinus armandi*
Chinese red pine	*Pinus tabulaeformis*
Japanese black pine	*Pinus thunbergii*
Larch	*Larix* spp.
China larch	*Larix principis-repprechtii*

Understory plants

Bamboos	*Pleioblastus* spp.
Bamboos	*Sasa* spp.
Hazelnut	*Corylus heterophylla*
Oriental thuja	*Platycladus orientalis*
Korean ginseng	*Panax ginseng*

Mixed Northern Hardwood Forests: Warm Temperate Forests

Trees

Oak	*Quercus*
Maple	*Acer*
Hornbeam	*Carpinus*
Alder	*Alnus*
Walnut	*Juglans*
Poplar	*Populus*
Hackberry	*Celtis*
Ash	*Fraxinus*
Basswood	*Tilia*

Understory plants

Dogwood	*Cornus*
Euonomys	*Euonomys*
Spicebush	*Lindera*
Dwarf gorse	*Ulex gallii*
Heather (true heather)	*Calluna vulgaris*
Heaths	*Erica* spp.

Animals of Eurasian Temperate Broadleaf Forests

Mammals

Golden snub-nosed monkey	*Rhinopithecus roxellana*
Rhesus monkey	*Macaca mulatta*
Raccoon dog	*Nyctereuctes procyonoides*
Giant panda	*Ailuropoda melanoleuca*

(Continued)

Himalayan black bear	*Selenarctos thibetanus*
Leopard	*Panther pardus*
Siberian tiger	*Panthera tigris*
Sable	*Martes zibellina*
Wild pig	*Sus scrofa*
Musk deer	*Moschus berezovskii*
Sika deer	*Cervus nippon*
Red deer	*Cervus elaphus*
Chinese goral	*Nemorhaedus caudatus*

Birds

	Haliaeetus albicilla
White-tailed Sea Eagle	
Golden Eagle	*Aquila chrysaetos*
Great Bustard	*Otis tarda*
Crested Shelduck	*Tadorna cristata*
Mandarin Duck	*Aix galericulata*
White-naped Crane	*Grus vipio*
Red-crowned Crane	*Grus japonensis*
Black Stork	*Ciconia nigra*
White-bellied Black Woodpecker	*Dryocopus javensis richardsi*
Fairy Pitta	*Pitta nympha*
Ring-necked Pheasant	*Phasianus colchicus torquatu*
Koklass Pheasant	*Pucrasia macrolopha*
Brown-eared Pheasant	*Crossoptilon mantchuricum*

EUROPE

Middle European Forests

Trees

European beech	*Fagus sylvaticus*
Pedunculate oak	*Quercus robur*
Sessile oak	*Quercus petraea*
Sycamore maple	*Acer pseudoplatanus*
Lime	*Tilia cordata*
European white birch	*Betula pendula*
Scots pine	*Pinus sylvestris*

Białowieża Forest and Central European Mixed Forest

European beech	*Fagus sylvaticus*
Pedunculate oak	*Quercus robur*
European hornbeam	*Carpinus betulus*
Lime	*Tilia cordata*
European alder	*Alnus glutinosa*
Black alder	*Alnus glutinosa*
Common ash	*Fraxinus excelsior*

Norway maple	*Acer platanoides*
European aspen	*Populus tremula*
Downy birch	*Betula pubescens*
Weeping birch	*Betula verrucosa*
Elm	*Ulmus scabra*
Common Elm	*Ulmus campestris*
Hornbeam	*Carpinus occidentalis*
Scots pine	*Pinus sylvestris*
Norway spruce	*Picea abies*
Wild apple	*Malus silvestris*
Purple willow	*Salix purpurea*

Understory plants

Dwarf birch	*Betula nana*
Cloudberry	*Rubus chamaemorus*
Willows	*Salix* spp.

Sub-Mediterranean Forests

Trees

Italian oak	*Quercus frainetto*
Downy oak	*Quercus pubescens*
Pyrenean oak	*Quercus pyrenaica*
Sessile oak	*Quercus petraea*
Pedunculate oak	*Quercus rober*
Tatarian maple	*Acer tatricum*
Ash-elm	*Fraximus ornus*
Elm	*Ulmus glabra*
Hophornbeam	*Ostrya carpinifolia*
Hornbeam	*Carpinus occidentalis*
Spanish chestnut	*Castanea sativa*
Limes or Lindens	*Tilia* spp.
Boxwood	*Buxus sempervirens*

Understory plants

Dwarf gorse	*Ulex gallii*
Calluna	*Calluna vulgaris*
Heaths (heather, etc.)	*Erica* spp.

Animals of European Temperate Broadleaf Deciduous Forests

Mammals

Western European hedgehog	*Erinaceus europaeus*
Eurasian red squirrel	*Sciurus vulgaris*
Long-tailed shrew	*Sorex mintus*

(Continued)

European shrew	*Sorex araneus*
Edible dormouse	*Glis glis*
Garden dormouse	*Elionys quercinus*
Red-backed or Bank vole	*Clethrionomys glareolus*
Yellow-necked mouse	*Apodemus flavicolis*
European hare	*Lepus europaeus*
Wood tarpan[a]	*Equus caballus*
Wild boar	*Sus scrofa*
Red deer	*Cervus elaphus*
Fallow deer	*Dama dama*
Roe deer	*Capreolus capreolus*
Wisent	*Bison bonasus*
Lynx	*Lynx lynx*
Eurasian badger	*Meles meles*
European polecat	*Mustela putorius*
Steppe polecat	*Mustela eversmanni*
Ermine	*Mustela erminea*
European mink	*Mustela lutreola*
Stone marten	*Martes foina*
Pine marten	*Martes marte*
Otter	*Lutra lutra*
Wolf	*Canis lupus*

Note: [a]Extinct.

Birds

White-tailed Eagle	*Haliaeetus albicilla*
Spotted Eagle	*Aquila clanga*
Garganey	*Anas querquedula*
Snipe	*Gallinago gallinago*
Kentish Plover	*Charadrius alexandrinus*
Spotted Crake	*Porzana porzana*
Black Grouse	*Tetrao tetrix*
White-backed Woodpecker	*Dendrocopus leucotos*
Green woodpecker	*Picus viridis*
Gray-headed Woodpecker	*Picus canus*
Old World Cuckoo	*Cuculus canorus*
Pied Flycatcher	*Ficedula hypoleuca*
Red-breasted Flycatcher	*Ficedula parva*
Nuthatch	*Sitta europaea*
Great Tit	*Parus major*
Blue Tit	*Parus caeruleus*
Blackbird	*Turdus merula*
Song Thrush	*Turdus philomelos*

Nightingale	*Luscinia megarhynchos*
Ortolan Bunting	*Emberiza hortulana*
Corn Bunting	*Miliaria calandra*
Savi's Warbler	*Locustella luscinioides*

SOUTH AMERICA

Trees

Monkey-puzzle tree	*Araucaria araucana*
Alerce	*Fitzroya cupressoides*
Chilean cedar (Cordilleran cypress)	*Austrocedrus chilensis*
Roble	*Nothofagus obliqua*
Hualo	*Nothofagus glauca*
Rauli	*Nothofagus alpina*
Lenga	*Nothofagus pumilio*
Ñirre	*Nothofagus antarctica*
Evergreen southern beech	*Nothofagus dombeyi*
Lingue (litchi)	*Persea lingue*
Avellano or Chilean hazelnut	*Gevuina avellana*
Chilean laurel or Chilean sassafras	*Laurelia sempervirens*
Olivillo or Palo muerto	*Aextoxicon punctatum*
Winter's bark	*Drimys winteri*
Ulmo	*Eucryphia cordifolia*

Understory plants

Bamboo	*Chusquea quila*

4

Mediterranean Woodland and Scrub Biome

The geographic regions of the world that are characterized by a mediterranean climate type exhibit a rather distinctive woodland and scrub vegetation that is associated with the dry summers and the mild and mostly humid winters of this climate. It is a climate pattern in which rainfall is concentrated during the relatively mild winter half of the year with precipitation levels that average 10–40 in (25–100 cm) per year. Very cold and prolonged freezing temperatures as well as snow are rare in this climate, except at high elevations. In all cases, these climate regions are on the western or southwestern coasts of continents, between 31° and 46° latitude in the Northern Hemisphere and between 28° and 42° latitude in the Southern Hemisphere. The Mediterranean Woodland and Scrub Biome takes its name from the region where it covers the greatest area: the Mediterranean Basin. Expressions of this biome are encountered in five widely separated regions on six continents: around the Mediterranean Sea in Europe, Southwest Asia, and northwestern Africa; in southwestern California; in west-central Chile; in the Western Cape province of the Republic of South Africa; and in southwestern and southern Australia (see Figure 4.1).

The geographic region of the Mediterranean Woodland and Scrub Biome in the Mediterranean Basin was where western science first took root. During the third and fourth centuries B.C., early Greek naturalists like Theophrastus and the perhaps better-known Aristotle were observing and recording the workings of nature in the Mediterranean Basin. It is from such early recorded observations that we have good scientific knowledge of the composition of the vegetation association, as well as description of the cause-and-effect of human actions in this region.

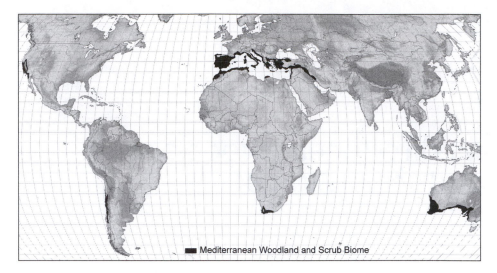

Figure 4.1 World distribution of Mediterranean Woodland and Scrub Biome. *(Map by Bernd Kuennecke.)*

When Europeans began to colonize the other mediterranean regions of the world, in North and South America, in South Africa, and in Australia, the quest for knowledge had sufficiently advanced and there were naturalists among the early explorers of these other regions. From these we have the descriptions of major exploratory expeditions, complete with relatively thorough inventories of native plants and animals that were encountered. A number of scientific studies on what is known today as convergent evolution were originally inspired by the apparent similarities in structure that were noted particularly among the plants found in the five mediterranean climate regions. This hypothesis holds that species of very different ancestry that are found in similar environments will over time evolve so that they will look and function much the same, having had to adapt to similar conditions. The biome has remained a focus of many modern comparative ecological and biogeographic studies.

Global Overview of the Biome

The shrubs in mediterranean regions display adaptations to drought, nutrient-poor soils, and fire. Most of the broadleaf plants in this biome have hard leathery leaves and are typically evergreen. Examples of such plants are live oaks and heaths (see Figure 4.2).

Significant disturbances of both natural and human-induced origin have had a tremendous impact on the development of the biome in all geographic regions. The perhaps most important impacts are frequent burning and livestock grazing. Burning is most commonly caused by lightning during the dry summer months. In spite of

Figure 4.2 Heaths of the Mediterranean Biome in South Africa. *(© by Susan L. Woodward, used by permission.)*

such alterations and influences over time, mediterranean ecosystems are home to about 20 percent of the world's plant species. This environment is one in which we find a rather high proportion of endemic species and genera, many of which are forms that have originated in the respective geographic regions of this biome. Many are found today only within this biome. Such exclusiveness is indeed a hallmark of mediterranean scrubland in all of its geographic expressions and is especially the case in the Western Cape of South Africa. It is because of such uniqueness that all five regions of the Mediterranean Woodland and Scrub Biome are listed among the world's 25 hotspots of biodiversity and endangerment. Hotspots are places with exceptionally high numbers of species, many of which are often not encountered elsewhere on Earth. Simultaneously, hotspots are environments in which the maintenance of high biodiversity is severely threatened by human activities.

One characteristic found throughout most geographic regions of the biome is that at least two distinct types of shrubland are encountered. The occurrence of a coastal shrubland type and an interior shrubland type is typical. Wherever mountains and predominant wind directions cause orographic effects (for example, the rainfall increases on the windward side), annual precipitation at the higher elevations may be sufficient to support a change from scrub to woodlands and even forests. With increasing latitude, rainfall totals and length of the rainy season also increase. As a result, there is a change from scrub to woodlands and forests in the higher latitudes of the expression as well as at the higher altitudes. Even where

·······························

Local Names for Mediterranean Scrub Vegetation

Each area of mediterranean scrub vegetation is located in a different linguistic region of the world. Subsequently, a number of local names for this vegetation type have emerged. Among such names we find the terms of *chaparral* and *coastal sage* in California. In Chile, the vegetation type is known as *matorral* and *espinal*. In Spain, it is referred to as *tomillares*; while the French call it most commonly *maquis*, although *garrigue* is locally used as well. The Italians refer to it as *macchia*, and the Greeks call it *phyrgana*. It is known as *bath'a* or *goresh* in Israel. In South Africa, the terms *fynbos* and *strandveld* are used, with the terminology originating from the former Dutch colonists and settlers of the area, while it is called *kwongan* and *mallee* in Australia with a nod to the terminology that originated with the aborigines of that continent.

·······························

woodlands and forests occur, however, a summer dry period remains. Scientists thus classify at least some of these woodlands as an integral part of the Mediterranean Woodland and Scrub Biome. When the trees of the woodlands and forests do not exhibit the typical adaptations to drought that mediterranean shrubs possess, then most scientists believe that those woodlands and forests should be left out of this biome. Therefore, the forests and woodlands that do occur as a result from higher annual rainfall levels will be identified in this chapter, but no full descriptions will be provided for them.

The geologic age of the surfaces supporting this biome is relatively young in North Africa, the Middle East, North America, and South America. These regions all have undergone recent mountain-building processes and thus have high mountains with significant local relief. As a result, they offer a great variety of microhabitats based on different combinations of sun exposure, elevation, type of bedrock, and slight differences in precipitation. The topographic variations, in particular, result in microclimatic differences and are reflected in the complex mosaic of plant communities common throughout these geographic areas. The regions of the biome in South Africa and Australia are markedly different in their geophysical characteristics. The biome occurs on geologically stable surfaces with minor local relief that is developed on the very ancient rock materials of old continental shields. The lack of recent mountain-building processes and the relatively long isolation of these regions from other geographic areas with subtropical vegetation is significant: It is the reason these two expressions of the biome have such high numbers of species in general and endemic species in particular.

Climate

The mediterranean climate—Cs in the Koeppen climate classification (see Table 1.1)—is rather unique among the global climate types. It is a climate that is characterized by the rainy season of the year coinciding with the cool or the winter period (see Figure 4.3). Winter is, of course, usually the nongrowing season for plants. Summers are usually dry. The total annual precipitation ranges between 15 and 40 in (380 and 1,000 mm) and varies depending on factors such as terrain, exposure to weather, and so forth. Elevation and exposure to prevailing winds have a significant

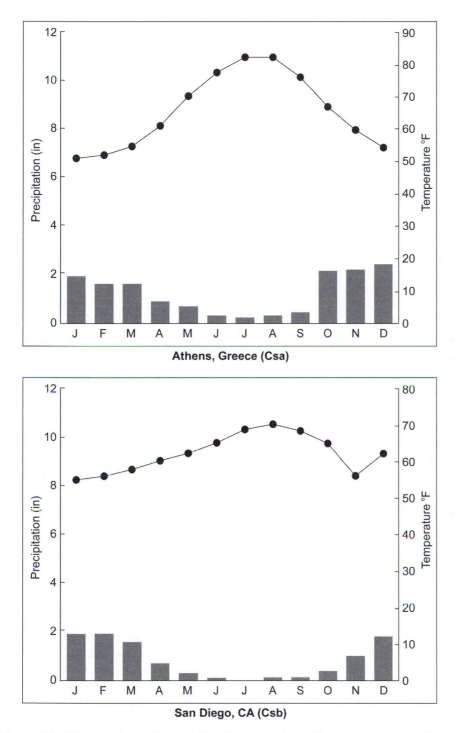

Figure 4.3 Climographs for Csa and Csb climates: Athens, Greece Csa, and San Diego, California Csb. *(Illustration by Jeff Dixon.)*

effect on precipitation levels. With increasing elevation, temperatures typically decrease at a rate of 2.8° F per 3,300 ft (5° C per 1,000 m), and lower temperatures mean a lower capacity of the air to hold moisture. The total amount of rainfall is enough to classify the mediterranean climate in the humid category. However, the unusual timing of the dry season, during at least part of the growing season, imposes significant stresses on plant and animal life for a few months each year. The stresses are similar to those experienced by plant and animal life in desert regions.

The mediterranean climate type displays a subtropical temperature pattern in which summers typically can be hot (unless moderated by elevation or high latitude), and winters normally are cool. Actual freezing temperatures are relatively rare. The proximity to the sea has a significantly moderating influence in all areas where this climate type occurs. During the winter months, the sea, which remains warmer than the continental land masses, usually prevents freezing temperatures. During the summer months, the fogs so often generated by the cool ocean currents flowing just offshore also have a moderating influence that makes extremely high temperatures a rare event. The combination of annual precipitation and temperature patterns results in a relatively limited but predictable period of time in which there is both sufficient soil moisture and ample warmth for active plant growth. Adaptations on the part of plants and animals must have occurred so that they can withstand these seasonal rhythms in climatic conditions and survive and thrive in the Mediterranean Woodland and Scrub Biome.

Soils

The soils of the Mediterranean Woodland and Scrub Biome developed on different continents with a great variety of geologic structures as parent material. No single soil type is characteristic of the mediterranean climate. One of the alfisol subgroup of soils, the xeralfs, are associated with regions that have mediterranean climate characteristics. These are naturally productive soils and are farmed intensively throughout the regions of their occurrence. They have little or no natural vegetation cover left. The xeralfs are, however, somewhat of an exception. Most soil types in this biome are deficient in the important plant nutrients nitrogen and phosphorus. In contrast, a few soil types may be rich in calcium, iron, sodium, and magnesium. Soils of the mountainous slopes of this biome are typically shallow, perhaps the result of long human use and the accelerated soil erosion associated with it. Many areas exhibit signs of low water infiltration rates in the form of rills and gullies cut by surface runoff. These soils may also have water-repellant properties, or they may be too compacted, or simply too stony, to accommodate easy water infiltration. Commonly found throughout this biome is one of the more fertile types of soils known as *terra rossa* (red earth). Terra rossa is a fine-grained red material that is not really a zonal soil as it does not reflect regional climatic conditions. Instead, the red material is rich in iron and calcium and is most usually found in areas with limestone as the bedrock. There is no real agreement among soil scientists as to the origin of the terra rossa soils. Some hold that it originated on limestone during a time in geologic history when the bedrock was exposed to an intense tropical climate and high humidity levels. Others think that terra rossa soils really are the result of aeolian transport and deposition and that the materials themselves originated outside the Mediterranean region and have been blown into their present locations by prevailing winds. Recurrent events supporting this idea are the infrequent dust storms that rise from the Sahara and blow northward. They deposit very fine silt-size materials as far north as the Italian Alps, where the reddish silt discolors the snow.

Vegetation

Throughout most geographic regions of the biome, the natural vegetation consists of evergreen scrub. It is not certain that evergreen scrub is indeed the only or even the primary response to the mediterranean climate. These regions have been inhabited by people, in most cases for long periods of time. This has often involved rather intensive use of the land. Numerous scientists firmly believe that woodlands and savannas at one time covered most mediterranean climate areas prior to the arrival of humans and especially the agricultural methods that people introduced. Cultivation and pastoralism altered the landscapes to a significant degree. The vegetation had to survive frequent burning and, at least in some regions, grazing and browsing by livestock. Human-induced changes favored the now-existing shrubs, which were not only able to withstand the annual summer droughts, but also were tolerant of thin and mostly nutrient-poor soils.

The plants of the Mediterranean Biome seem well adapted to the climatic patterns of their regions. Most of the woody plants have tough, evergreen, sclerophyll leaves that resist dehydration. Scientists see this as an adaptation to the short growing season and prolonged summer drought. Plants that maintain foliage on a year-round basis do not lose precious time producing new leaves at the beginning of each brief growing season. Although that may be a growth advantage, the same plants had to simultaneously develop the necessary protection for maintaining foliage throughout the dry conditions of the annual summer droughts. Leaves need to be shielded from drying out. Some common adaptations shared by the evergreen plants throughout the Mediterranean Biome are designed to reduce evapotranspiration and to prevent dehydration—for example, they have developed leathery leaves with waxy cuticles and characteristically deeply sunken stomata (that is, the pores through which all leaf tissue of plants exchange gases, such as water vapor and carbon dioxide, with the atmosphere). In addition, the leaves of some plants have developed either upright or vertical orientation to reduce the damaging direct exposure of leaf tissue to the heat from sunlight. Other plants have adapted to reduce evaporation rates and leaf surface temperatures by developing particular surface colors, such as a reflective silver color, or they have developed hairy leaves, and some have both.

Some broadleaf plants' adaptation to heat and drought is expressed by needle-shaped leaves that result in a reduction of leaf surface areas that helps to reduce water loss. Yet other plants have adapted by becoming "drought-dimorphic." In such cases, plants possess two types of leaves. During winter and the growing season of spring, they have larger and softer leaves; with the onset of drier conditions in late spring and early summer, these are replaced with smaller and harder leaves for the summer drought period. Other plants are "drought-deciduous," which means that they shed their leaves at the onset of the dry summer months.

Mediterranean shrubs typically have both long tap roots to reach water that is stored at depths and an extensive network of roots near the surface that permits them to intercept water from rainfall as soon as it percolates into the soil. Many of the nonwoody plants of this biome have water storage tissues such as bulbs and tubers or succulent leaves and stems. This enables them to survive the drought seasons. Numerous plants in the Mediterranean Woodland and Scrub Biome, the so-called annuals and ephemerals, begin and complete their life cycles in a single growing season. Their adaptation to drought is that they survive as seeds protected beneath the forest litter or in the upper soil horizon.

Symbiotic relationships between many plants and fungi enable the growth of plants in the mediterranean biome. Soil conditions are typified by low-nutrient content in the root zone. An association with fungi growing around their roots helps plants gain access to limited nutrients—that is, the fungi grow long, thin filaments into the soil to extract the minerals that are important to plant growth but not accessible to the roots of the plants. The fungi are able to convert these minerals into compounds of nitrogen, phosphorus, and potassium that plants can use. They exchange these compounds with the plants' roots for some of the carbon compounds that plants produce in photosynthesis. The association involving the exchange of compounds between plants and fungi is called a mycorrhiza. The fungi are commonly referred to as mycorrhizal fungi. Mycorrhizae are not limited to the Mediterranean Scrub Biome. Such associations between plants and fungi are indeed common throughout all biomes; however, the survival of many species included in the Mediterranean Scrub Biome depends on this association. Similar processes are performed by species of the protea family (Protaeceae) in Australia and Africa. Proteas form fine, cottony rootlets near the ground surface at the beginning of the rainy season. This lets them extract the nutrients that leach from the decaying forest litter with the onset of rains. After several months, these rootlets will wither and die. Some grasses, sedges, and restios (reed-like plants of the family Restionidaceae) have similar structures that enable them to access nutrients that otherwise would not be available to their normal roots.

Fire is a common occurrence in the regions of the world that have a mediterranean climate type. The summer drought period leaves the living biomass and the litter dry; it is also commonly a time of heat lightning. The regions that have experienced long-standing human occupation have been subjected to fires set by humans for a variety of purposes. As a result, fire has had an extremely important role in natural and human-dominated disturbances in the Mediterranean Woodland and Scrub Biome. Fires eliminated those plants that were intolerant of frequent or regular burning. Some plants, however, have developed ways to survive fires, such as an extra thick tree bark that acts as a protective insulation blanket against quick ground fires. Some trees in this biome even require burning to be able to regenerate. Burns may destroy the surface tissue of some plants, but many mediterranean shrubs are often referred to as "sprouters," because they are quite capable of resprouting from root crowns, the thickened woody structures at the base of the trunks of these plants. A prime example of a sprouter is chamise (see Plates XII and XIII), the characteristic shrub of the southern California chaparral. Other plants cope with the impact of fire in a different way. The so-called seeders are able to rapidly recolonize a burned-over site through their production of large quantities of seeds. Their seeds often have the ability to remain viable in the litter or in the top layer of the soil horizons for many years, even decades, until a fire occurs. The heat from the fire then stimulates the germination of these seeds. One example from the Northern Hemisphere regions of the biome are the closed-cone pines, such as the Monterey pine, endemic to California. Unopened cones remain on their branches until there is a fire. The cones open after being exposed to heat and the seeds are released. The fire has left the ground newly enriched with a layer of ash, so that the released seeds find fertile soils to germinate in. A similar process is common among some of the native plants in the Southern Hemisphere areas of the biome in Australia and in South Africa.

Perennial plants that have renewal buds well below the ground surface are called geophytes. They are especially abundant in the Mediterranean Woodland and Scrub Biome and demonstrate a high diversity. They survive fires because the soil basically insulates the bulbs, corms, or rhizomes from the heat of the average wildfire at and above the soil surface. Geophytes quite often demonstrate their survival through the production of a colorful wildflower display in the nutrient-enriched postburn sites in the spring after a fire. Excellent examples of such displays after wildfires are the brodiaeas and mariposa lilies in California.

A final characteristic property to mention is that many of the plants endemic to this biome are aromatic. The Mediterranean Basin is the region of origin for perfumery and culinary herbs such as oregano, sage, thyme, and lavender. These plants have aromatic oils in their leaves that may offer several different advantages assisting their survival. Aromatic oils often render the foliage unpalatable to many herbivores. Reduction in predation is the result and contributes to the survival of the plants to reproductive age. The oils, on the other hand, contribute to the flammability of the plants and thus may actually promote the rapidity and intensity of fires. This may give such plants a competitive edge over other plants that are less well adapted to survival in areas with frequent fire. Additional evidence shows that these aromatic oils actually help plants survive better during times of intense heat from solar radiation. The partial evaporation of these oils fills the stomata and thus reduces the loss of water vapor that would otherwise occur through the leaf pores.

Animal Life

The animal life of the Mediterranean Woodland and Scrub Biome has no representatives that are exclusively "Mediterranean." A number of animal lifeforms have the ability to tolerate or to avoid drought and heat. Most animals have a relatively high mobility. Many can go to free water sources on a daily basis and drink at springs or from permanent streams. Being mobile also means that they can avoid the exposure to high heat during the day by moving into a shaded spot or burrowing below ground where it is cooler. They can shift the active part of their daily life to nighttime when temperatures are lower. Some animals of mediterranean regions have physiological adaptations to withstand the drought season and the heat. These adaptation are similar to those we encounter among the animals of the desert regions. Some may simply enter a state of torpor or dormancy (aestivation) during the summer drought months. Other animals utilize their mobility and seasonally migrate to higher elevations or latitudes with lower summertime temperatures. As in desert regions, a few animals in the mediterranean regions are capable of surviving on the water that they can extract from their food. Those animals do not require drinking water at all and can survive as long as their food sources are available.

Each of the world's mediterranean regions is somewhat isolated in a geographic sense, separated by oceans, by mountains, and deserts. Such isolation has favored the evolution of a number of animals that are encountered in only a single expression of this biome, just as isolation has done for plants.

MAJOR REGIONAL EXPRESSIONS OF THE MEDITERRANEAN WOODLAND AND SCRUB BIOME

The Mediterranean Basin: Maquis and Garrigue

The French terms *maquis* and *garrigue* will be used throughout the following discussion for the two main types of scrub vegetation that we find in the Mediterranean Basin, regardless of the country in which this vegetation is encountered (see Figure 4.1). The Mediterranean Basin is where the biome reaches its maximum geographic extent; its physical size is three times the combined area of the other four regions. It serves here as the model against which we will compare the other regional expressions of the biome. The scrub vegetation of the Mediterranean region is not continuous, but fragmented into a number of different sections, each of which is separated from the others either by mountains or by the main body of the Mediterranean Sea and its many extensions along the Apennine and the Balkan peninsulas. Maquis and garrigue vegetation is typically found in Europe at elevations below 3,300 ft (1,000 m). We find this vegetation on the Iberian Peninsula, in southeastern France, on the Apennine Peninsula (Italy), on the Balkan Peninsula, and on the islands of the Mediterranean Sea from the Balearic Islands in the western part to Cyprus in the eastern part. On the eastern shores of the Mediterranean Sea, in Southwest Asia, mediterranean scrub is found in Turkey, Syria, Lebanon, Israel, and western Jordan. On the southern shores of this sea in North Africa, the biome is encountered only in the Maghreb region, which is located (from west to east) in the northern parts of Morocco and Algeria and northwestern Tunisia. Throughout this large region are many genera, species, and subspecies restricted to only small parts of this region and absent from other sections. Throughout the northern parts of the Mediterranean Basin, significant variations in the number of actual species present in any given area of the region occur. The highest number of species is found near the center of the Mediterranean Basin, on the Balkan Peninsula.

Tall shrubs and short trees with varying canopy heights characteristically form the association of woody plants of the maquis (see Figure 4.4). Evergreen trees with hard leaves include the holm oak, the Kermes oak, the strawberry tree, the carob, the mastic tree or lentisk, and the wild olive. The needleleaved junipers and the Aleppo pine are widely distributed. Among the tall shrubs are several species of rock rose as well as oleander. Commonly attaining heights of 12–20 ft (4–6 m), these shrubs often form a rather dense evergreen woodland quite different from any found in the other regions of the Mediterranean Woodland and Scrub Biome.

The maquis contains a rather large number of bulbs, annuals, and ephemerals considered to a large extent to be the result of human activities. They may be the long-term result of interactions with agricultural and pastoral systems that have existed within this region for more than 5,000 years. The practices of plow cultivation, as well as keeping land open for grazing, create on a regular basis relatively short-lived open patches of ground. Annual and ephemeral plants have evolved to take advantage of such open sites. The light weight of their wind-blown seeds enables them to quickly and easily invade such sites, especially if, for any reason, the intensity of human activities has slackened. The seeds of these plants typically only germinate in full sun. The plants, too, require high solar exposure to thrive and are not at all shade tolerant. Because of this shade intolerance, once woody mature

250 ft

Figure 4.4 Vegetation profile: maquis in the Mediterranean Basin. *(Illustration by Jeff Dixon.)*

plants shade such clearing or open sites, the annuals and ephemerals are forced to disperse to other or new open sites so that they may survive. They have become quite adept in this process of continuous invasion of open sites. When we encounter such plants as transplants in other environments throughout the world, we find them to be highly successful weeds.

The second common scrub type of the Mediterranean Basin is garrigue. It consists typically of low shrubs on sites that are characteristically much hotter and drier than those that support maquis. Soils on limestone are often relatively dry due to the high porosity of such bedrock. Consequently, garrigue is commonly associated with hard limestone bedrock throughout the region. The shrubs of the garrigue are typically either spiny or aromatic. A substantial portion of the aromatic shrubs are members of the mint family, such as lavender, thyme, rosemary, and sage. Drought-dimorphic and drought-deciduous shrubs are common. The high diversity of geophytes found here is second only to the Western Cape of South Africa. Within this section of the biome, many native terrestrial orchids occur, as well as numerous species of tulip, iris, narcissus, crocus, and cyclamen. In Greece, this low shrub vegetation is commonly referred to as phyrgana. In Israel, it is called bath'a. Throughout much of the region, scientists consider the garrigue to be a degraded form of maquis. Human misuse of the land is probably at the root of such degradation of the plant communities.

The animal life, just like plants, has been subjected to significant human modifications and impacts over time. A number of large mammals inhabit this region, although some survive only in remote areas. Included are herbivores such as the fallow deer, the roe deer, the European wild boar, and the ibex. Native carnivores include the Spanish lynx and the wolf, both of which occur only in remote areas and are today under protection. The biome does include among the larger mammals one primate, the Barbary macaque. It is found in southern Spain (and is under particular protection in Gibraltar), as well as in North Africa. Among a number of smaller mammals are insectivores such as hedgehogs, shrews, and moles. Endemic rodents include some voles and spiny mice. Of particular interest is the world's smallest mammal, the Etruscan shrew, which is endemic in this region. With a weight of only 0.05–0.08 oz (0.6–2.4 g) it is indeed tiny.

Throughout the Mediterranean Basin, there are significantly more reptiles than amphibians. Considering the climatic characteristics, that is not surprising, since reptiles by their very nature are preadapted to the summer drought of this region. A variety of lizards, tortoises, snakes, and two chameleons are associated with this region.

Most resident birds are widely distributed throughout the biome, unlike many of the other animals as well as plants. The diversity of birds is not high and relatively few are endemic. Many of the birds, however, have differentiated into subspecies, as, for example, in the cases of the Blue Tit and the Eurasian Jay. The most impressive aspect of birdlife in this expression of the biome is the large number of nonresident species present throughout much of the year. The Mediterranean Sea and the land areas along its shores are major stopover and staging sites for vast numbers of migrating birds, particularly for birds that breed in Central, Northern Europe, and northeastern Europe. Scientists have estimated that between 230 and 250 Eurasian species (thrushes, wrens, kinglets, and so on) spend their breeding season outside the Mediterranean Basin, but return for the nonbreeding season. This results in a situation in which the number of bird species in the Mediterranean Basin is significantly higher in winter than in summer. In addition, another 130 species utilize the Mediterranean Basin as a stopover place twice per year in their annual migrations between Eurasian breeding grounds and their wintering areas in sub-Saharan regions. The total number of birds that cross the Mediterranean Sea during spring and again in fall has been estimated by one scientist to be 5 billion. A bird-of-prey has even adjusted its breeding season to the migrations of other birds: the bird-eating Eleonora's Falcon nests on cliffs along the Mediterranean shoreline and breeds in late summer and early fall, which is a rare breeding time for birds. However, this assures the availability of a food supply at a time when the falcon needs to feed its young.

Western North America: Chaparral and Coastal Sage

The mediterranean climate type in North America and the associated expression of the Mediterranean Woodland and Scrub Biome occur along the western coast of the North American continent from San Francisco (37° 45′ N) to northern Baja California, Mexico (31° N) (see Figure 4.1). Of the four ecoregions identified as having mediterranean conditions in California three are part of this biome: the California Interior Chaparral and Woodlands, the California Montane Chaparral and Woodlands, and the California Coastal Sage and Chaparral. (The fourth ecoregion is the California Central Valley Grasslands.) It is an expression of the biome that is restricted in its geographic extent by the mountain ranges that run mostly parallel to the coastline. Within this narrow belt, we encounter two distinctly different plant associations, chaparral (from the Spanish *chapa,* scrub oak) and coastal sage. Chaparral plants consist of a rather rich variety of hard-leafed evergreen shrubs commonly growing in a mosaic of plant communities that reflect the recent fire history of each site. It is a region in which the plant communities are typically subjected to fire in a cycle that seems to repeat itself every 12 to 20 years. Repeated fires at such frequencies help to perpetuate a community of chamise (see Plate XII). The needle-like small leaves of this shrub contain highly flammable oils that assist in increasing fire intensities. While fire consumes the aboveground portions of the plant, chamise is quite capable of rapidly resprouting from root crowns that are shielded from the fire by the soil (see Plate XIII). In addition, chamise is highly tolerant of dry and low-nutrient soil conditions. Because of these abilities, chamise occurs rather frequently in nearly pure stands that have a limited ground or herb layer. It is one of the most widespread plants in the Californian mediterranean formation and indeed is considered the indicator species for the biome in this geographic region (see Figure 4.5).

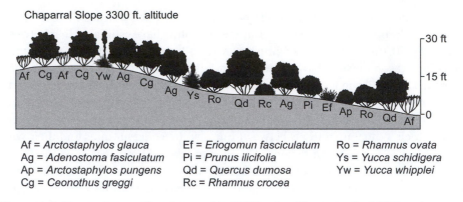

Chaparral Slope 3300 ft. altitude

Af = *Arctostaphylos glauca*   Ef = *Eriogomun fasciculatum*   Ro = *Rhamnus ovata*
Ag = *Adenostoma fasiculatum*   Pi = *Prunus ilicifolia*   Ys = *Yucca schidigera*
Ap = *Arctostaphylos pungens*   Qd = *Quercus dumosa*   Yw = *Yucca whipplei*
Cg = *Ceonothus greggi*   Rc = *Rhamnus crocea*

Figure 4.5 Vegetation profile: chaparral in California. *(Illustration by Jeff Dixon.)*

Fire is the element that favors the dominance of chamise. Wherever fire is less frequent, we encounter a shrub diversity that is surprisingly high. On such unburned sites, a plant association develops that consists of mountain mahogany, sumac, California lilac, toyon, scrub oak, and manzanita. The canopy height of such diverse vegetation reaches 8–12 ft (2.5–4 m), depending on the specific site conditions. Most plants in such stands develop from seeds, rather than by resprouting from root crowns. Within such areas closed-cone pine, such as knobcone pine, may appear in the association. When fires have occurred and killed the shrubs within a particular stand, sites are initially repopulated by bulbs and annuals that may form a virtual carpet of wildflowers. Larkspurs, brodiaeas, mariposa lily, and poppies burst into a rainbow of colors. Such developments are, however, relatively short-lived. As the shrub components of the plant association become reestablished, the nonwoody species will be subjected to severe competition and will decline in numbers. Whenever stands are on north-facing slopes, which typically are more humid (less evaporation due to lower solar exposure), a forest will develop of the chaparral shrub and trees, such as California bay or laurel, live (hard-leafed, evergreen) oaks, Pacific madrone, and golden chinkapin. At the upper-elevational levels of the chaparral, temperatures decrease and precipitation levels increase due to altitude, and trees such as the Coulter pine and the endemic big-cone Douglas-fir are mixed in the chaparral that is dominated by manzanita.

The past decades have been a time during which humans have not only caused fires, but, just as important, have prevented wildfires or at least have prevented their extensive spreading. Whenever grazing, browsing, and chaparral fires have been suppressed or prevented for at least 50 years, such as we find on parts of Santa Cruz and Catalina islands (parts of the Channel Islands west of Los Angeles), a so-called elfin forest of live oaks develops. Some scientists think that if such a fire suppression continues over even longer periods of time, then an oak savanna would develop as the climatically determined natural vegetation type. Some think that oak savanna would cover much of the inland areas with a mediterranean climate type in North

America if it were not for fire. On the mainland of California, several areas with a mediterranean climate type are indeed vegetated by several oaks, some evergreen and some deciduous, in open grassland situations. These exceptions give credence to what might happen if fire were suppressed for long periods of time in this environment.

Coastal sage is the second mediterranean vegetation type on the West Coast of North America. The chief indicator for this vegetation type is the presence of soft-leafed low shrubs. Coastal Sage is usually found directly along the coast, rarely extending more than a few miles inland. It is a vegetation that is highly influenced by the humidity derived from the sea fog that forms over the cool California Current off shore. The vegetation typically has an open canopy so that there is sufficient sunlight to encourage an abundance of forbs and grasses to develop in the ground layer. In general the soft-leafed shrubs in this expression of the biome are less tolerant of drought conditions than the hard-leafed types commonly found inland. Many coastal sage plants are drought-dimorphic or drought-deciduous. Included among the dominant species of the coastal sage are California sagebrush, purple sage, white sage, Munz's sage, toyon, and black sage.

Which forests along the California coast should be included in the Mediterranean biome is a question that relates to some of the coastal forests to the north of San Francisco. Several authors include the stands of coastal redwoods. These giant conifers, as well as others on this section of the Pacific Coast, such as Douglas fir and western hemlock, grow in areas of high rainfall. In the case of the redwood stands, the precipitation is the result of an uplifting of airmasses coming off the Pacific Ocean that is caused by the presence of the Coastal Ranges and that produces high amounts of rainfall on the windward side of these mountains. High rainfall and nearly daily fog off the ocean produce sufficient moisture to eliminate the summer drought that would be necessary for the climate type to be classified as Mediterranean. The forests dominated by these giant conifers along the northern coast of California typically have an understory consisting of heaths (family Ericaceae): rhododendrons, azaleas, huckleberries, and related species. In this book, these forests are described in more detail under the Boreal Forest Biome (see Chapter 2).

Fire and Chaparral in California

Chaparral as an association of plants typically is a dense vegetation consisting largely of scrub oaks and other drought-resistant plants, many of which have waxy leaves as well as natural oils. The vegetation density, the summer-drought climate, and the flammability of the materials render this vegetation association highly susceptible to wildfires, and indeed, wildfires are common in these areas. Many of the plant species in this association seem to require fire as a cue for subsequent germination of seeds or opening of seed cones. However, no evidence indicates these plants actually have adapted to a particular regimen of fire. Indeed, it appears that the seeds of many plants in this association require the accumulation of a significant number of years of leaf litter (30 years or so) before they actually can germinate. Species that fall into this category include scrub oak and the holly-leafed cherry. When the interval between fires is significantly shortened, chaparral is often replaced with grassland. Mature chaparral areas that burn after many years of litter accumulation typically are characterized by all-consuming crown fires. Reinvasion by native species and ecological succession are necessary on the burned-over site before chaparral becomes reestablished.

Ecological Islands

The southern California coastal region has a number of so-called ecological islands that have endemic conifers with extremely restricted ranges of occurrence. One of these is the Monterey pine, a closed-cone pine found naturally at only three sites. The total land area of these sites is about 11,000 acres (4,250 ha). The irony may be that, in its home range, the tree is limited in terms of its total geographic extent. This tree has been widely introduced as a fast-growing timber crop in other parts of the world. It grows well in other environments and has been considered a scourge that is leading to the demise of native trees in these new growing areas. Other conifers that are endemic to the coastal region of California include Monterey cypress, Gowen cypress, and the Torrey pine.

Animal life in the Mediterranean Biome of North America is characterized by a high diversity of species. This is an area with a patchwork of habitats, the result of microclimatic differences forged by its position along a coast and the presence of mountain ranges to its east. Habitat variety allows for a variety of wildlife. Large herbivores, such as mule deer and elk, are found wherever habitat is suitable. Among predators are the coyote, the kit fox, the mountain lion, and the bobcat. A number of small mammals are common in this expression of the biome, such as the endemic white-eared pocket mouse, the San Diego pocket mouse, kangaroo rats, and numerous other rodents. The Mediterranean Woodland and Scrub Biome of southern California is home to some 100 species and subspecies of birds such as Acorn Woodpecker and California Scrub-jay. Among them is the endangered California Gnatcatcher, a bird that is found only in the coastal sage area. Another endemic bird, not often seen, is the secretive Wrentit. It prefers dense chaparral as its habitat.

Amphibians and reptiles are represented by many species. The lizard genus *Liolaemus* dominates with its 34 species. Five endemic or nearly endemic species of lungless salamanders (family Plethodontidae) also occur. This expression of the biome is rather well known for being home to many different butterflies and a high diversity of native bees.

Western South America: Matorral and Encinal

In Central Chile, the mediterranean scrub vegetation region is a near-perfect mirror image to that of the North American mediterranean region (see Figure 4.1). In South America, the biome lies between the Pacific Ocean and the Andean Cordillera. It extends from La Serene (31° S) south to Conception (37° S). The cold Humboldt or Peru Current flows northward offshore. This current and its temperature under westerly winds generates significant coastal fogs that are known in Chile as *camanchacas*. Similar to other mountainous regions where we encounter an expression of this biome, the mountainous topography, associated vertical microclimatic changes, and the latitudinal extent of the biome create a variety of microhabitats. The Mediterranean Woodland and Scrub Biome is locally known as matorral (from the Spanish *mata*, shrub). Typically it is a patchwork of many different plant communities with diverse structural characteristics and a variety of species in the respective patches. The South American matorral has many more deciduous plants than the California chaparral. When we compare the two expressions further, we find that the Chilean part of the Mediterranean scrub biome also has many more species with thorns than what is encountered in its northern counterpart. Tall columnar cactuses are prominent among the thorny plants. Cacti or other succulents are nearly completely absent in the other expressions of the Mediterranean Scrub Biome. They are also a strong indicator that fire is not an important element in the succession of plant communities in the matorral.

In South America, the coastal matorral is comparable to the coastal sage of California. It is also somewhat similar to the garrigue in the Mediterranean Basin as well as to the strandveld of the Western Cape, South Africa (see below). The coastal matorral consists of a vegetation characterized by low and soft-leafed shrubs (see Figure 4.6). Many of the plants are drought-deciduous. One example is the wild coastal fuchsia. In contrast to its counterparts in the garrigue and the strandveld, the plant association of the coastal matorral contains tall terrestrial bromeliads or puyas. These all have sharp, serrated leaves. And a columnar cactus grows here as well.

Away from the coast, in the interior, the thorny *espino* is by far the most dominant shrub. Another common plant in the canopy layer is mesquite. This vegetation is typically referred to as "espinal" to distinguish it from other types of matorral. In these stands, mesquites, acacias, and other shrubs, such as chilca, huañil, and trevo, reach heights of 6–20 ft (2–6 m). They grow in an open, savanna-like thicket, with

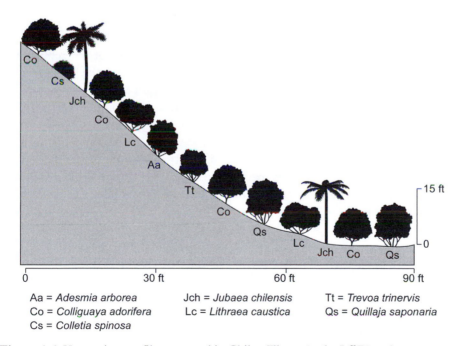

Aa = *Adesmia arborea*   Jch = *Jubaea chilensis*   Tt = *Trevoa trinervis*
Co = *Colliguaya adorifera*   Lc = *Lithraea caustica*   Qs = *Quillaja saponaria*
Cs = *Colletia spinosa*

Figure 4.6 Vegetation profile: mattoral in Chile. *(Illustration by Jeff Dixon.)*

scattered puyas and cacti. The ground between shrubs is covered mostly with perennial bunch grasses. Vines and bulbs are also common, as is a ground layer of spring-blooming ephemerals. Wherever the moisture levels increase on south-facing slopes, the open shrub canopy common on the drier sites changes to a closed canopy. Some indications are that espinal may be the result of overgrazing and woodcutting in the area. The latter was clearly aimed at harvesting usable woods and nearly completely eliminated the hard-leafed trees such as litre and quillay that once grew here.

The slopes of the coastal ranges of central Chile are the geographic regions where hard-leafed evergreen woodlands and forests can be found today. One of the most common trees is peumo. It is in the same family (Lauraceae) as the bay laurel of the Mediterranean Basin and the California bay of North America. Here also is the Chilean palm, the world's southernmost palm. However, this palm is a rapidly disappearing component of the plant association of this forest. Toward higher elevations or farther south in areas where rainfall is more plentiful during the winter rainy season, southern beeches dominate the forest stands. Hard-leafed evergreens typically occur at the lower elevations; at the higher elevations, where more moisture precipitates from airmasses being forced upward, deciduous species grow. A common understory plant in *Nothofagus* forests is bamboo (see also Chapter 3). Still farther south or higher in elevation are cooler temperatures and even more precipitation from winter. Stands of an endemic conifer, the monkey-puzzle tree, grow in such areas.

The Mediterranean Woodland and Scrub Biome of Central Chile is not known for having a diverse animal life. Nearly half of the 26 amphibians are unique to the region. One is the pointy-nosed Darwin's frog, one of two species in the family Rhinodermatidae. Reptiles are a more diverse group with 39 species, most of which are lizards. A large proportion are endemic: one-third of the reptiles are found nowhere else on Earth. The matorral is habitat for relatively few native mammals, among them are the guanaco and the Chilean fox, as well as several rodents. The majority of the small mammals, such as the South American field mice, are omnivores. Only the leaf-eared mouse and the rice rat feed nearly exclusively on seeds. One small mammal, the mouse marsupial, is insectivorous. A total of 175 birds occur, not a particularly great diversity. Some are of significance as seed dispersers for the plants of the mediterranean shrubs. These so-called fruit-eaters or frugivores include birds such as doves, thrushes, the Chilean Mockingbird, and the Diuca Finch.

Western Cape, South Africa: Fynbos and Strandveld

On the southern tip of Africa (see Figure 4.1) in the Western Cape province of the Republic of South Africa, is a region of extremely high species richness and such a large number of endemic plants that biogeographers place the flora of the region in a Kingdom of its own. It is typically referred to as the Cape Floristic province. The local name for this expression of the Mediterranean Woodland and Scrub Biome is "fynbos," a term that means "fine bush" and refers to the fine-leafed plants that make up about 80 percent of the Cape Floral Kingdom. It is a biome with more than 8,600 species (68 percent of which are endemic) in more than 200 genera, of which 198 are endemic. The relative isolation of this region from others is in part responsible for the fact that so many plant species in this region are found nowhere else on earth and that six plant families are endemic to this biome. Among the main plant families represented in the fynbos are the proteas or sugarbushes (family Protaceae) (see Figure 4.7); 69 of these families are endemic. Among them are plants such as the common protea, the snow protea, the king protea, pincushions, and common sunshine cone bush. A second characteristic and endemic family is that of the reed-like shrubby grasses known as "restios" (family Restionaceae) (see Plate XIV). A third important plant group is the widely distributed heath family (Ericaceae), which includes species such as fire erica, sticky rose heath, and mealie heath (see Figure 4.2 and Plate XV). Bulbs (geophytes) are conspicuous members of the fynbos plant community and include members of the iris family (Iridaceae), including gladiolus, spotted African corn lily, the large wild iris, and freesia; the amaryllis family (Amaryllidaceae), including belladonna lily and bush lily or clivia; and the arum family (Araceae), represented by the calla lily. Many of these bulbs are familiar garden and house plants throughout the Northern Hemisphere today.

Well represented and highly diversified in this region is the daisy family (Asteraceae). A similarly high diversity and presence is shown by the orchid family

Figure 4.7 Protea in South Africa. (© *Susan L. Woodward, used by permission.*)

(Orchidaceae). Other members of the plant formation of the fynbos are cycads, which are ancient cone-bearing plants that may reach sufficient size to be referred to as trees. Although they have a rather distant relationship to pines, they superficially look like palms.

The fire-prone fynbos is mostly characterized by evergreen shrubs that reach varying heights and produce several vertical layers in the vegetation. The spring months in the Southern Hemisphere are September and October, when there is a significant peak in the flowering of many plants. The bloom period extends nearly year-round in the eastern parts of the region that receive more precipitation and thus have slightly moister soils than the western fynbos. The year-round blooming of some species is attributed to the fact that many of the species are either tropical or subtropical in their origins. The heath family with some 3,000 species constitutes most of the plants. They are typically low shrubs, attaining heights of 2–6 ft (0.5–2 m). Many of the proteas (see Plate XVI) have showy flowers and normally vary in height from 6–12 ft (2–4 m). Reaching a greater height than the heaths, they often provide a dense overstory. The majority of the plants in the fynbos regenerate from seed. There are some exceptions: a few of the proteas will resprout after fire from root crowns. There is only one real tree species in the fynbos, the silvertree. It typically can be found only on the more humid lower slopes of Table Mountain near Capetown.

In the areas of the Mediterranean Woodland and Scrub Biome along the coast of South Africa, the vegetation is commonly known as strandveld and is typically dominated by evergreen restios. Other plants include low-growing shrubs, succulents, and bulbs. The Cape Floristic province is a region that contains the world's greatest diversity of geophytes and most of these are to be found in the strandveld of the Republic of South Africa.

Some scientists consider another association of shrubs in South Africa a third formation of the Mediterranean Woodland and Scrub Biome. This is the renosterveld, a vegetation dominated by low-growing, mostly fire-prone, small- and soft-leafed shrubs of the daisy family. Most notable among them is the renosterbos or rhinoceros bush. Renosterveld is encountered on the relatively fertile soils that are derived from shale formations in the coastal lowlands.

Animal life is relatively diverse in the Mediterranean Woodland and Scrub Biome of the Western Cape of Africa. Insects play an important role in plant propagation and are thus important elements of the animal life of this region. In the fynbos, plants are pollinated mostly by flying insects; moths and some beetles, in particular, have evolved close symbiotic relationships with specific plant species. Harvester ants are important seed dispersers. Scientists have estimated that there are about 1,200 plant species in this expression of the biome that depend entirely upon ants collecting and burying their seeds. The adaptation on the part of some plants has progressed to a point that the seeds of ant-dispersed plants have globules of fat attached to them to attract ants. The ants take the seeds to their underground nests, where they are stored and protected from fire and predators.

Some 109 different reptiles and amphibians inhabit this geographic region. They include endemic chameleons and numerous tortoises. Of special mention is the Cape ghost frog, a frog with a large mouth when a tadpole. Its suckered toes permit it to attach itself to rocks as an adult in the fast-flowing water of streams. In this manner, it can graze the algae growing there.

Birds of all types are rather common and many are endemic. Cape Sugarbirds, seeking nectar in proteas, pollinate the plants. The same pollinating functions are carried out for aloes as well as proteas by sunbirds, the South African equivalent of the New World Hummingbirds. Mammals are numerous and diverse as well. Small antelopes like the Cape grysbok, the common duiker, the steenbok, and the klipspringer easily hide in the shrubs. Among the larger antelopes are the relatively rare bontebok and the Vaal rhebok. Chacma baboons are populous, even in developed areas. Carnivores include leopards, which still occasionally prey on the endemic herbivores.

Like the plant life of the Western Cape, the animal life includes many forms that are not found elsewhere on Earth. Some are evolutionarily rather primitive. One example is the Cape ghost frog that was mentioned above. Such primitive forms attest to the long isolation of this region. The relative geographic isolation, until recently, has contributed to the ability of more primitive forms to prevent their displacement by more advanced forms that originated in other parts of the world.

Western and Southern Australia: Kwongan and Mallee

The Mediterranean Woodland and Scrub Biome occurs in Australia in two separate geographic regions. The first is located in the State of Western Australia, while the other is in the State of South Australia (see Figure 4.1). A physical separation of the two regions exists in the form of the arid Nullarbor Plain. The Western Australian region is extremely rich in terms of number of species; indeed, it has the highest number of species of any geographic areas in all of Australia. Within this region are hundreds of trees and shrubs in the genera *Acacia* and *Eucalyptus* alone. Only about 30 species in each of these genera occur in the South Australia expression of the biome. The local term kwongan derives from the southwestern Australia aboriginal language and is a word for open, scrubby vegetation that occurs on sandy soils. A second used in conjunction with this vegetation is mallee, which is an aboriginal word that refers to the growthform of those eucalyptus trees capable of resprouting from root crowns and developing into multistemmed shrubs.

Kwongan resembles the inland scrub formations of other mediterranean regions, particularly those of South Africa's fynbos. Kwongan commonly consists of a number shrubs—commonly prickly moses and small trees of the genus *Banksia* (family Protaceae). These may grow up to a height of 20 ft (6 m) if there is a sufficiently long time between the reoccurring fires in the area. A number of other proteas in the kwongan are typically shrubs. Most regenerate by seed after fires have occurred. Some plants, such as desert banksia, beaked hakea, and the sheoak *Casu-arina muellerana*, actually store their seeds on their branches in woody capsules that will only open to release the seeds after fires have occurred. Other plants drop their seeds quite normally, but their seeds have a thick coating that has to be removed by a ground fire. In their dormant state, these seeds can actually wait for a period of several years for a fire to remove the tough coating, which helps the germination process. The ground layer of this association is species rich as well. Typically a carpet of color in spring comes from blooming strawflowers or everlastings, as well as from annuals in the daisy family, common in this layer. Characteristic for this association also are sundews (family Droseraceae); 47 species are known from the vicinity of Perth alone. These insectivorous plants utilize insects as a source of nitrogen to help them overcome the deficiencies of low-nutrient soils. Several other common plants that form characteristic components of the kwongan include the grass tree, restios, and shrubs of the Australian heath family (Epacridaceae). Also occurring are members of the citrus family (Rutaceae), the fanflower family (Goodeniaceae), and the kangaroo paw family (Haemodoraceae).

The mallee association of plants is heavily dominated by the often pungent, gray-green evergreen shrubs of the genus *Eucalyptus*. The trait of resprouting from root crowns common among these eucalypts has enabled them to survive the frequent fires of the area. Locally known as mallee heath, a rather diverse understory of heath-like shrubs (family Epacridaceae) is encountered on the poorest soils. When soils are somewhat richer in terms of nutrients, an understory consists of grasses and forbs. The highly flammable porcupine grass is a well-known member of the understory in the drier parts of the South Australian region. This grass typically grows in hummocks that have a diameter of nearly 3 ft (1 m) and leaves that end in sharp spines that can puncture the skin of a passing animal or human quite easily, just like the quills of a porcupine.

Karri forests are prevalent in the wettest areas in the southwestern tip of Western Australia. The hard-leafed karri trees grow to enormous sizes, attaining heights of 250 ft (76 m). Below the canopy, a multilayered understory of smaller trees, saplings, shrubs, and vines develops in such a forest. These forests are home to a large number of terrestrial orchids. Where sites are drier, forests consist of nearly pure stands of jarrah. The hard-leafed jarrah trees can reach heights of about 100 ft (30 m). These drier jarrah forests typically have an understory layer consisting of shrubs that include banksias, sheoaks, and paperbarks.

The species richness found among plants is not duplicated for animals. However, animals play important roles as pollinators and seed disperses in both the kwongan and the mallee. Insects, birds, and mammals are all involved. The pollination of the proteas is facilitated by nectar-feeding birds, among them the honey-eaters (family Meliphagidae). This region is home to the endemic honey possum, a tiny nocturnal marsupial. Other animals pollinating members of the protea family include bats.

Additional animal inhabitants of the kwongan and the mallee are the western gray kangaroo, as well as the rather rare and unique numbat. The latter is a small termite-eating marsupial anteater. It is also the state mammal for Western Australia. Another rare and small marsupial is the quokka. Small wallabies and hare wallabies, both of which are grazers and browsers, also inhabit the biome, as do rat kangaroos and bandicoots. This is the region where the bush rat is found, one of Australia's few native, nonmarsupial mammals. It feeds on bulbs and other geophytes. Another unusual animal is the ground-dwelling Mallee Fowl. It incubates its eggs in a heap of composting leaf litter.

Reptiles are diverse and numerous, as is the case for Australia as a whole. The southwestern region alone has more than 190 recognized species, of which 26 percent are endemic. Among them are several tortoises. Amphibians are less diverse than reptiles, but nearly 80 percent of Australia's amphibians are restricted to those areas that have a vegetation cover of mediterranean scrub.

Human Impacts

Humans have influenced, altered, and eradicated portions of the Mediterranean Woodland and Scrub Biome in all regions where it occurs. In most cases such influences have been rather extensive; in some cases they reach back to great antiquity. People have affected the vegetation by starting frequent fires, overgrazing, removing the original vegetation for agriculture, harvesting wood for timber and other wood products (primarily charcoal), applying a variety of agricultural practices, introducing nonendemic species, converting the land to urban uses, and developing transportation corridors. In the past few decades, additional impacts have resulted from the increasingly important role of tourism and the associated residential and infrastructure development. All of these impacts have had a role in shaping the vegetation associations we encounter in these regions today. It cannot be denied that humans have fundamentally changed the landscapes of the regions of the world that had developed plant associations adapted to the mediterranean climate.

The Mediterranean Basin has been and continues to be one of the most densely settled regions of the inhabited world. Very little is left of the natural vegetation, and what is left is only on highly inaccessible sites with poor soils and slopes too steep for human land uses. People have utilized fire as a tool for manipulating natural vegetation for at least 100,000 years. They have probably done so in this region for most if not all of that time. The eastern sections of the Mediterranean Basin were some of the cradle areas of agriculture. Here are the places of origin of many agricultural plants, such as rye, oats, wheat, barley, peas, lentils, grapes, olives, and so on and of domestic animals, such as cattle, pigs, sheep, and goats. Domestication processes began at least 10,000 years ago. The first evidence of forest clearing to provide land for farming activities dates back some 8,000 years. About 5,000 to 4,000 years ago agricultural activities were broadly distributed throughout this vast region. Agriculture set the stage for another process in this region: the development of the urban cultures of Western civilization. Classical literature provides us with glimpses of what the region might have been like during the Classical Greek Period. Homer reports that the region was forested with live oaks, pines, cedars, wild carob, and wild olives. Harvesting of the famed cedars of Lebanon began 5,000 years ago. The classical time period was not without critics to such destruction of natural environmental factors. In fact, quite some concern was expressed about the effects of deforestation first by Greek and later by Roman writers. During these periods of great population growth, forests increasingly were eradicated to clear land for agriculture and provide timber for construction of dwellings in the growing towns and to build water crafts for both trade and warfare. Whenever hillsides became denuded of their natural vegetation, severe erosion typically followed, at times exposing the bedrock and leaving in most areas only a thin veneer of soil far too shallow to allow the redevelopment of protective vegetative covers. Only the open scrub was able to survive under such conditions. Humans then utilized such open scrub vegetation for grazing and browsing by sheep and goats so that a livelihood could be derived from these poor lands. Livestock exerted new pressures on the vegetation. Plants that were unpalatable for goats and sheep had the advantage and were able to colonize the dry, open, low-nutrient surfaces between the shrubs. Primarily, these shrubs were annuals and "weeds." Much of the contemporary maquis and garrigue is the result of human land use over the past few thousand years. Scientists believe that the hard-leafed evergreen scrub originally was confined to the coastal zones of the region. With increasing human-caused disturbances, these plants expanded their aerial extent since they were more adapted to the new situation. Of course, human impacts in this region (as well as in some of the other Mediterranean regions) were significantly assisted by natural factors, such as naturally steep slopes, and by the fact that rainfall often occurs in high volume over short periods of time, resulting in sheet flow and flash floods. The region has frequent earthquakes and a relatively weak subsurface geology (particularly the existence of soft marls or highly jointed and fragmented beckrock), all of which combine to enhance the severity of human impacts on the natural plant associations of the region.

While the early agricultural developments and the later classical time period certainly were times when the landscapes of the Mediterranean Basin changed dramatically, not all the effects of human activities are old. Quite the contrary, many contemporary developments are changing the landscapes once again. The post–World War II period brought economic changes to the region. The European part of the Mediterranean Basin has witnessed a rural depopulation since the 1960s. The changes brought about by the agricultural policies of the European Common Market and the European Community have caused populations to gravitate to urban and industrial centers and away from farming. In the European part of the basin, the problems of overgrazing of the past have been replaced with problems of undergrazing. Decreased grazing has allowed a dense cover of flammable shrubs to develop. This, in turn, has greatly increased both the severity and the frequency of fires.

The situation in the African part of the basin is quite different. Since the countries of the Maghreb of North Africa gained their political independence in the late

1960s, the region has experienced an increase in rural populations and an associated explosion in the number of livestock being grazed on the seminatural vegetation cover. In all areas of the Mediterranean Basin, contemporary urban development is expanding, bringing with it a destruction of the surrounding shrublands as well as conversions of formerly agricultural land to urban land uses. Coastal habitats throughout the region are rapidly being converted to residential and tourist resorts in the wake of the continuing emergence of the economic importance of tourism. The space-consuming elements that accompany resort development include infrastructure (particularly modern roads with their wide right-of-way), private estates, parks (public and private), marinas, beachfront access, and so forth. All of these elements mean eradication of natural the vegetation, often accompanied by replacement vegetation that is imported for aesthetic reasons.

Upon arriving in the other mediterranean climate regions of the world, European settlers were struck by the similarities to the Mediterranean Basin. The apparent similarities in natural vegetation prompted European settlers to grow mediterranean crops, so that these distant regions are now used for growing grapes, olives, citrus fruits, figs, and wheat.

Human impacts in the mediterranean region of North America certainly predate the arrival of European settlers in California. Aboriginal peoples practiced burning in conjunction with numerous activities, including hunting and agriculture. With great probability this had expanded the area that was originally covered by chaparral. The arrival of European settlers greatly accelerated this conversion. Cattle grazing was widely practiced during the Spanish, Mexican, and early American periods of the settlement history of the region. It initiated major land degradation and conversion of the former vegetation associations. One consequence of grazing practices was the replacement of native annual and perennial bunch grasses with European sod-forming annuals such as brome and wild oats (see Figure 4.8).

The fire suppression policies that have been instituted during modern times in California have had a significant impact on the nature of the vegetation and on the fires and their characteristics. By controlling or preventing fires year after year, the composition of species in the plant association rapidly changed. At the same time, fuel accumulated in form of both the living shrubs and the litter. When a fire does occur after several years of suppression, the intensity is often high as a result of such fuel accumulations, and fire infernos will at times rage out of control and destroy the chaparral rather than assist in the natural renewal of this vegetation association. Low intensity or "cool" fires typically occur when there is less dense scrub and where there are low volumes of accumulated litter as is typical for areas with frequent ground fires. "Cool" fires will not alter the soil chemistry but may actually increase the solubility of various compounds, thus providing a more nutrient-rich environment in which the vegetation association can become reestablished by the regrowth of both sprouters and seeders. Whenever "hot" fires occur following long periods of fire suppression, they typically alter the soil structure, reducing the porosity of the soil surface, and even contributing to the development of nearly

Figure 4.8 Goat grazing damaged the chaparral vegetation of Catalina Island, California. (© *Susan L. Woodward, used by permission.*)

water-repellant soil surfaces. The majority of the seeds stored in the ground litter cannot survive temperatures in excess of 300° F (150° C). Since many hot fires exceed that limit, such fires actually can prevent the rapid recolonization by seeders that would stabilize the soils on slopes and thus increase the possibility of accelerated soil erosion. The aftermath of hot fires often is mudslides, which develop on such denuded sites with the onset of the winter rains. The modern encroachment of suburbia onto these hazard-prone slopes often brings with it high costs in terms of property damage and occasionally the loss of human life when fires sweep through the wooded subdivisions and estates that have been established in the chaparral.

Agricultural activities continue to reduce the geographic extent of the chaparral. As in the other mediterranean regions, the California region is also a major vegetable-, fruit-, wine-, and cut flower-producing region. The moisture deficiencies of the summers are being overcome with modern irrigation practices. In the course of expanding agricultural lands, natural vegetation is removed. The use of heavy farm machinery and equipment, the applications of fertilizers and pesticides, the mechanical impacts of loosening the soil surface layers, and the utilization of irrigation technologies have long-term impacts on the soil structure and chemistry of this region. Such soil alterations may actually prevent the reestablishment of the natural chaparral when fields and orchards are abandoned. The problems of salinization are realized in some areas, and toxic levels of selenium have accumulated as a result of some irrigation practices. The combined impact of modern human activities on the coastal sage shrub are estimated to have reduced the total area of this vegetation association to less than 10 percent of its pre-Columbian extent.

The matorral of the mediterranean region of Chile may have experienced some adverse influences from its aboriginal peoples. After all, they did practice the herding of llamas, tree-cutting for firewood, and irrigation agriculture. However, scientists seem to relegate the beginnings of severe human impacts on the matorral to the onset of European settlements. Spanish explorers and colonists brought cattle and other livestock to Chile. The initial grazing lands for cattle were natural meadows, which quickly proved insufficient in size for growing cattle populations. To create more pastureland, the vegetation surrounding natural meadows was burned. Overgrazing resulted in a degradation of pastureland, and cattle were replaced first with sheep and later with goats. Goats proved to be the animals best capable of sustaining themselves on shrubs while providing milk, meat, fiber, and leather to the settlers. The abundance of thorny plants distinguishes the Chilean matorral from other Mediterranean scrub. The presence of grazing and browsing livestock may have contributed rather strongly to the emergence of thorny shrubs as dominants in the matorral. The relatively low frequency of natural fires may be another factor that favored thorny shrubs. Early in the twentieth century, European rabbits were introduced into Chile. Their grazing activities may have contributed to the current dominance in the matorral of thorny shrub. Rabbits seem to have caused a marked decline in bunchgrasses and the seedlings of woody species except where such vegetation is protected under thorny brush.

The espinal of Chile may have its origin in human disturbances. Tree-cutting for fuel and charcoal certainly has contributed to the demise of trees. Harvesting of other wood products caused further impacts. The bark of the quillay, known as soapwood in the English-speaking world, develops a gentle lather in water. It was harvested and exported to clean fine fabrics, lenses, and precision instruments in the wealthier portions of the world. Excessive harvesting of wood also occurred in regions where copper mining industries flourished during the nineteenth and twentieth centuries. Charcoal was used for smelting the ore.

With the Spanish conquest, the region witnessed the introduction of European agricultural crops and practices. Initially these efforts seemed to concentrate on the clearing of land to provide fields for the cultivation of wheat. Dry-farming techniques, which involved the shifting of cultivation from one field to another, were practiced. Large areas of matorral were destroyed when this practice was first introduced. The rate of destruction increased during the nineteenth century, resulting from expansion of dry-farming areas in response to market demands for wheat associated at that time with gold rushes in California and Australia. Toward the end of that century, a lucrative market for the relatively dry wheat from Chile's mediterranean region was developed in Britain, causing further increases in dry-farming areas.

The dry-farming practices used in Chile involved the clearing of large tracts of land, plowing the soil, and growing wheat, barley, cumin, peas, and chick peas for a period of three or four seasons. The continued use of the land exhausted both existing soil moisture and nutrients. Fields with decreased yields were abandoned and the soil was exposed to wind erosion, which created the landscape of an agri-desert. Over time, some shrubs have reinvaded the former fields. However, plant diversity on such reestablished stands is low. Furthermore, the invading plants seem to grow in straight lines parallel to each other and reveal still today the furrows and ridges made years ago.

The mediterranean climate region of Chile remains of high importance to that country's economy, since it encompasses a major region from which agricultural and wood products are exported. Sections of the matorral continue to be converted to orchards, vineyards, and commercial tree plantations. Trees from other mediterranean regions of the world are commonly preferred over native species. These nonendemic plants, chief among them Monterey pine and eucalyptus, especially blue gum (*Eucalyptus globulus*), are vastly preferred by landowners because they grow more aggressively. In addition, they are generally free of the diseases that might plague them in their native distribution areas.

In the mediterranean climate regions of South Africa's Western Cape, humans and our near ancestors have been a part of the local ecosystem for at least a million years. What has been called "fire-stick agriculture" has been ongoing for at least 100,000 years in this region and was practiced by ancient hunter-gatherers and the modern Khoi-San alike. It involves burning the vegetation so that edible geophytes are encouraged to grow, along with a new growth of grasses that might attract wild game to the burn areas. Nearly 2,000 years ago, the Khoi-Khoi entered this region, pushing the previous occupants, the San, into the more marginal habitats on mountain slopes, where the San continued the practice of regular burnings. The Khoi-Khoi brought with them a cattle-based pastoral economy; they were, of course, interested in encouraging grasses and other plants palatable to cattle to grow. Today some evidence in the form of pollen analysis indicates that both the hunter-gatherers (San) and the herders (Khoi-Khoi) intentionally influenced the vegetation of the Western Cape and increased the relative abundance of certain proteas as well as geophytes and grasses.

The Western Cape region of South Africa experienced its first European settlement in 1652. This date signaled the beginning of the introduction of agricultural practices that are fundamentally the same as in the Mediterranean Basin of Europe—vineyards, orchards, olive groves, and wheat fields—and, of course, associated with them the clearing of the native vegetation. This process is not completed at the present time. The Western Cape continues to be an important and even expanding agricultural region, with perhaps the newest addition to the variety of crops being native proteas for the cut flower industry. The introduction of European-type agriculture along with intensive cultivation has had the greatest of all human impacts on the native vegetation. The replacement of native vegetation by cropland is the most intensive alteration. As a result of past and present landscape conversions, the renosterveld has been nearly completely destroyed. Only remnant sites that have native or seminatural vegetation associations are left. Not only has the natural vegetation been affected, but also animal life has been strongly affected.

The extinction of several large mammals, among them the endemic bluebuck, the quagga, and the Cape lion, is the result of hunting.

The fynbos has been a tremendously species-rich expression of the biome. Its remnants are today threatened by residential and infrastructural expansions from growing populations and urbanization developments, as well as the impacts from forest replacement when afforestation is carried out with nonnative species. The natural vegetation is increasingly being invaded by alien species, such as those that have escaped from tree plantations. Among the invasive nonnative species are the Monterey pine from California, the maritime pine, and the Aleppo pine from the Mediterranean Basin to the north, and several plants imported from Australia. Problems are occurring not only with imported plants, but also with introduced animals. The Argentine ant could over time replace the native harvester ants. Native proteas depend on the harvester ants that store proteas seeds underground, but the Argentine ant leaves seeds on the surface where they can then be easily found by seed-eating animals. The potential loss of these native ants that have a symbiotic relationship with many native plants could change the abundance of many of the native shrubs in the fynbos of the Western Cape of South Africa.

Human impacts on natural vegetation in Australia date back at least 40,000 years. Evidence points to aboriginal people using fire on a regular basis to burn vegetation in association with hunting activities, although fires also occurred naturally from lightning. Scientists think that the impact of such fires upon species composition and vegetation structure was significant, although the exact extent of such influences is still being debated. The changes that resulted once Europeans arrived are well documented and have involved both impacts on the natural vegetation and impacts on the aboriginal peoples, who continued their way of life of hunters and gatherers until well after the Europeans had arrived. The human influences on the vegetation in Australia have been severe and, for the most part, are relatively recent: most occurred during the twentieth century. The European settlement of Australia began in 1829. The soils of western and southern Australia were too poor to attract agricultural settlers and were avoided until after the introduction of superphosphate fertilizers in the 1890s. Only with the assistance of such chemicals could the soils of the mediterranean regions of Australia be utilized for modern agriculture. Once these applications had become available, agriculturalists began to clear extensive areas in western and southern Australia of their natural vegetation and to cultivate wheat or establish pasture crops.

The introduction of sheep, cattle, and goats brought about a rapid expansion of the populations of these domesticated animals, and with it, the overgrazing of the native grasses. Farmers replaced the natives with imported forage species to support their growing herds of cattle and sheep. The conversion of natural vegetation to agriculture often resulted in desertification and salinization of streams and soils. Wild animals from Europe were also introduced, at times with negative consequences. Perhaps the best-known impacts of this type are the rabbit plagues that originated with the deliberate introduction of the European rabbit in 1859. This plague is not yet completely under control. The biggest problem associated with rabbits is that their overgrazing of natural vegetation has significantly accelerated surface erosion and increased desertification processes. The introduction of cats and other small carnivores has affected populations of native birds and marsupials, often important pollinators and seed dispersers of the native plants.

A plant disease known as "jarrah dieback" is a relatively modern plague in Australia that is related to the soil fungus *Phytophthora cinammoni*. It is believed that this fungus was introduced from Indonesia in the 1920s. It attacks roots in such a manner that they are subsequently unable to absorb water and water-soluble nutrients. The tall jarrah trees are most affected by this fungus and show a slow dieback of their crowns in the dry forests. While the impact on the jarrah trees is perhaps obvious, in part due to the size of the trees, the fungus affects and kills some 900 other plants as well. Most of the plants attacked by this fungus are natives. It seems that the fungus spreads rather slowly on its own in undisturbed forest soils. However, its spores spread rapidly when forest soils are disturbed by modern machinery or when forest soil is transported on lumber trucks or in conjunction with road construction activities. The fungus seems to spread rapidly when there is little or no forest litter, thus fire—a commonly used management tool used throughout Australia—seems to be another major factor in its spreading. In addition, burning has a tendency to suppress or eliminate prickly mosses, which repels the fungus. At the same time, fires seem to increase the occurrence of banksias, which do not act as a repellant to this fungus. The "dieback" that is observed in conjunction with this fungus is considered as *the* greatest threat to the native plants of the Mediterranean Woodland and Scrub Biome in Australia.

The kwongan and the mallee of Australia are experiencing an increasing presence of a number of additional nonnative plants that are mostly garden escapes. These include bulbs such as gladiolus, ixia, and freesia, all native to South Africa. Other garden forbs, such as the bristly European annual borage known as Paterson's curse and the South African daisy, are also common invaders.

Further Readings

Low, A. Barrie, and A. G. Rebelo, eds. 1996. *Vegetation of South Africa, Lesotho, and Swaziland*. Pretoria: Department of Environmental Affairs and Tourism.

Mittermeier, Russell A., Patricio Robles Gil, Michael Hoffmann, John Pilgrim, Thomas Brooks, Cristina Goettsch Mittermeier, John Lamoreux, and Gustavo A. B. Da Fonseca. 2004. *Hotspots Revisited*. Mexico City: CEMEX.

Walker Bay Fynbos Conservancy. 2001. http://www.fynbos.co.za.

Appendix

Selected Plant and Animals of the Mediterranean Woodland and Scrub Forest Biome (arranged geographically)

Mediterranean Basin Maquis and Garrigue

Maquis

Trees

Aleppo pine	*Pinus halepensis*
Carob	*Ceratonia siliqua*
Holm oak	*Quercus ilex*
Kermes oak	*Quercus coccifera*
Mastic tree or Lentisk	*Pistacia lentiscus*
Strawberry tree	*Arbutus unedo*
Wild olive	*Olea europa*

Shrubs

Oleander	*Nerium oleander*
Rock rose	*Cistus* spp.

Garrigue

Shrubs

Lavender	*Lavandula* spp.
Rosemary	*Rosmarinus officinalis*
Sage	*Salvia* spp.
Thyme	*Thymus vulgaris*

Mammals

Herbivores

Barbary macaque	*Macaca sykvanus*
European wild boar	*Sus scrofa*

(*Continued*)

Fallow deer	*Dama dama*
Ibex	*Capra hispanica*
Roe deer	*Capreolus capreolus*
Spiny mice	*Acomys* spp.
Voles	*Pitymys* spp.

Carnivores

Wolf	*Canis lupus*
Spanish lynx	*Felis pardina*
Etruscan shrew	*Suncus etruscus*
Eurasian hedgehog	*Erinaceus europaeus*

Birds

Blue Tit	*Parus caeruleus*
Eleonora's Falcon	*Falco eleonora*
Eurasian Jay	*Garrulus glandarius*

California Chaparral and Coastal Sage

Trees

Redwoods	*Sequoia sempervirens*
Gowen cypress	*Cypressus goveniana*
Monterey cypress	*Cypressus macrocarpus*
Douglas fir	*Pseudotsuga menzeisii*
Bigcone Douglas fir	*Pseudotsuga macrocarpa*
Coulter pine	*Pinus coulteri*
Knobcone pine	*Pinus attenuata*
Monterey pine	*Pinus radiata*
Torrey pine	*Pinus torreyana*
Western hemlock	*Tsuga heterophylla*
Pacific madrone	*Arbutus menziesii*
Golden chinkapin	*Castanopsis chrysophylla*

Chaparral

Shrubs

California bay or Laurel	*Umbellularia californica*
California lilac	*Ceanothus* spp.
Chamise	*Adenostoma fasciculatum*
Red shanks	*Adenostoma sparsifolium*
Manzanita	*Arctostaphylos* spp.
Scrub oak	*Quercus dumosa* and *Q. berberidifolia*
Gambel oak	*Quercus gambelii*
Bear oak	*Quercus ilicifolia*
Sumac	*Rhus ovata*

Toyon	*Heteromeles arbutifolia*
Mountain mahogany	*Cercocarpus* spp.
California coffeeberry	*Rhamnus californica*
Hollyleaf cherry	*Prunus ilicifolia*
Laurel sumac	*Malosma laurina*
Yucca	*Yucca whipplei*

Perennial forbs

Brodiaeas	*Brodiaea* spp.
Mariposa lily	*Calochortus* spp.
Poppies	*Rhomneys coulteri*
Larkspurs	*Delphinium* spp.

Coastal Sage

Black sage	*Salvia mellifera*
Purple sage	*Salvia apiana*
White sage	*Salvia apiana*
Munz's sage	*Salvia munzii*
California sagebrush	*Artemisia californica*
Toyon	*Heteromeles arbutifolia*

Mammals

Herbivores

Elk	*Cervus elaphus*
Mule deer	*Odocoileus hemionus*
Kangaroo rats	*Dipodomys* spp.
Sonoma chipmunk	*Eutamias sonomae*
White-eared pocket mouse	*Perognathus alticolis*
San Diego pocket mouse	*Ambrosia pumila*

Carnivores

Bobcat	*Felis rufus*
Coyote	*Canis latrans*
Kit fox	*Vulpes macrotis*
Mountain lion	*Felis concolor*
Bobcat	*Lynx rufus*

Birds

Acorn Woodpecker	*Melanerpes formicivorus*
California Gnatcatcher	*Poliotila californica californica*
California Scrub-jay	*Aphelocoa californica*
Wrentit	*Chamaea fasciata*

(Continued)

South American Mattoral and Encino

Trees

Chilean palm	*Jubaea chilensis*
Monkey puzzle tree	*Araucaria araucana*
Southern beeches	*Nothofagus* spp.
Peumo	*Cryptocarya alba*

Shrubs of the Matorral

Chilca or Seepwillow	*Baccharis glutinosa*
Coastal fuchsia	*Fuchsia lycoides*
Huañil	*Proustia pungens*
Litre	*Lithraea caustica*
Mesquite	*Prosopis chilensis*
Quillay	*Quillega saponica*
Thorny espino	*Acacia caven*
Trevo	*Dasyphyllum diacanthoides*
Bamboo	*Chusquea quila*

Succulents

Columnar cactus	*Trichocereus chilensis*
Puyas	*Puya berteroniana, Puya chilensis*

Mammals

Herbivores

Mouse marsupial	*Marmosa elegans*
Guanaco	*Lama guanicoe*
Leaf-eared mouse	*Phyllotis darwini*
Rice rat	*Oryzomys longicaudatus*
South American field mice	*Akodon* spp.

Carnivore

Chilean fox or Zorro	*Pseudalopex griseus*

Birds

Dove	*Columba araunac*
Diuca finch	*Diuca diuca*
Chilean mockingbird or Tenca	*Mimus thenca*
Austral Thrush	*Turdus falklandii*

Amphibian

Darwin's frog	*Rhinoderma darwinii*

South African Fynbos

Tree

Cycads	*Encephalartos* spp.
Silvertree	*Leucadendron argenteum*

Shrubs and Groundcover

Rhinoceros bush	*Elytropappus rhinocertis*
Bush lilly	*Clivia miniata*
Belladonna lilly	*Amaryllis belladona*
Calla lilly	*Zantedischia aethiopica*
Spotted African corn lily	*Ixia maculata*
Freesia	*Freesia alba*
Large wild iris	*Dietis grandiflora*
Gladioli	*Gladiolus* spp.
King protea	*Protea cyanoroides*
Common protea	*Protea caffra*
Snow protea	*Protea cryophila*
Sugarbush	*Protea repens*
Common sunshine conebush	*Leucadendron salignum*
Pincushion	*Leucospermum* spp.
Fire erica	*Erica cerinthoides*
Sticky rose heath	*Erica decora*
Mealie heath	*Erica patersonia*

Mammals

Herbivores

Bontebok	*Damaliscus dorcas dorcas*
Cape grysbok	*Raphicerus melanotis*
Common duiker	*Sylvicapra grimmia*
Klipspringer	*Oreotragus oreotragus*
Steenbok	*Raphicerus campestris*
Vaal rhebok	*Pelea capreolus*
Rock hyrax	*Procavia capensis*

Omnivores

Chacma baboon	*Papio ursinus*
Honey badger	*Mellivora capensis*
Aardvark	*Orycteropus after*

(Continued)

Carnivore

Leopard	*Panthera pardus*

Birds

Cape Sugarbird	*Promerops cafer*
Malachite Sunbird	*Nectarinia famosa*
Orange-breasted Sunbird	*Anthobaphes violacea*
Orange-breasted Sunbird	*Nectarinia violacea*
Forest Canary	*Serinus scotops*
Victorin's Warbler	*Bradipterus victorini*
Cape Rock-jumper	*Chaetops frenatus*
Cape Siskin	*Serinus totta*
Cape Sugarbird	*Promerops cafer*

Reptiles

Chameleons	*Bradypodium* spp.

Amphibian

Cape ghost frog	*Heleophryne rosei*

Western Australia Kwongan and Mallee

Trees

Grass tree	*Xanthorrhoea* spp.
Jarrah	*Eucalyptus marginata*
Karri	*Eucalyptus diversicolor*
Paperbarks	*Melaleuca* spp.
Sheoaks	*Casuarina* spp.

Shrubs

Prickly moses	*Acacia pulchella*
Desert banksia	*Banksia ornata*
Beaked hakea	*Hakea rostrata*

Grass

Porcupine grass	*Triodia scariosa*

Mammals

Marsupials

Numbat or Banded anteater	*Myrmecobius fasciatus*
Western gray kangaroo	*Macropus fulginosus*

Wallabies	*Macropus* spp.
Hare wallabies	*Lagostrophus* spp.
Quokka	*Setonix brachyurus*
Honey possum	*Tarsipes rostratus*
Rat kangaroo	
Bandicoot	

Placental mammal

Bush rat	*Rattus fuscipes*

Bird

Mallee Fowl	*Leipoa ocellata*

Glossary

Abscission. Controlled separation of a leaf from the branch upon which it grew.

Affinity. Relationship; is used in plant and animal geography to indicate taxonomic similarities between the flora and fauna of different regions.

Alien (species). Nonnative; living outside its natural range. Often referred to as introduced or exotic species; term used with plants.

Anaerobic. Living or being without free oxygen.

Annual. Plant that completes its life cycle in either one year or one growing season.

Arboreal. Living in trees; adapted to life in the trees.

Asexual Reproduction. The multiplication of individual organisms by processes that do not involve the combining of ova and pollen (plants) or egg and sperm (animals). Examples include budding and cloning. See also vegetative reproduction.

Base Cations. The most weak and prevalent acid cations in the soil materials.

Biodiversity. The total variability of life contained in genes, species, and ecosystems.

Biogeography. Geographic distribution of life on Earth, past and present, and the processes that determine where plants, animals, and other organisms occur.

Biome. Large geographic region with similar vegetation, animal life, and environmental conditions; one of the largest recognizable ecological units on Earth.

Broadleaved. Plants with thin, flat leaves; contrast to needleleaved.

Canopy. Uppermost layer of foliage in a vegetation association.

Climate. Average weather (especially temperature and precipitation) patterns that occur during a normal year and are experienced over decades or centuries. Weather refers to the conditions of the atmosphere at any given moment.

Climax Community. Stable, persistent plant community established in response to the regional environmental conditions; the theoretical end stage in ecological succession; often referred to as Climax Vegetation.

175

Closed Canopy. Crowns of adjacent plants in the same layer of the vegetation inter-mingle and overlap to prevent much sunlight to penetrate to layers below the canopy.

Community. The living organisms (that is, species) assembled in a given area; refers at times only to a subset of these species, such as the plant community, or the bird community, or the benthic community.

Compound Leaf. Leaf composed of several distinct leaflets.

Conifer. Cone-bearing plant such as the needleleafed pines, spruces, hemlocks, firs, and so on.

Cover. Proportion of a surface on which vegetation occurs, usually measured as a percent.

Crown. Top mass of foliage on a tree or shrub.

Cushion plant. Low-growing, multistemmed plant that grows as a dense mound; growthform is mostly associated with cold and dry climate regions.

Deciduous. Plants, usually trees and shrubs that shed their leaves during the nongrow-ing season.

Desertification. Process of degrading the quality of plant cover and soils as a result of human overuse with plants and animals.

Dispersal. Movement of an organism away from its place of origin; refers to the move-ment and establishment of a species beyond the limits of its previous distribution area.

Disturbance. Event or process that disrupts an ecosystem; usually destroys at least part of the ecosystem affected.

Drift (glacial). Glacial deposits, both sorted and unsorted; includes outwash and till.

Drought-dimorphic. Change in plant morphology to compensate for drought season.

Drought-evading Foliage. Plants are drought-deciduous and drop most or all of their leaves during summer drought.

Drumlins. Depositional landform consisting of glacial till; oriented in the direction of movement of the ice.

Ecology. Science that studies the interrelationships between organisms and their environments.

Ecoregion. Relatively large area of land (or water) that contains a geographically dis-tinct assemblage of natural communities that share a large majority of their species, dynamics, and environmental conditions.

Ecosystem. Association of living organisms and their physical environment with which they interact.

Ecotone. Transition zone between two adjoining ecosystems or biomes that may vary in width depending on localized conditions.

Eluviation. Process in soils in which there is a downward transport of fine soil particles from upper to lower soil horizons.

Endemic. Originating in and restricted to a particular geographic area or region.

Eolian. Wind-caused erosion, transport, and deposition of materials.

Ephemeral. Plant that completes its life cycle in an extremely short period of time (usu-ally a few weeks).

Epiphyte. Plant that grows on the branches or stems of another plant, using its host for support only.

Ericaceous. Reference to a member of the plant family Ericaceae (includes heathers, rhododendrons, bilberries, Labrador tea, and so on).

Eskers. Depositional landform that develops when glacial meltwater has carved a tunnel under the ice that becomes filled with meltwater-carried rock debris.

Evapotranspiration. Merging of the processes of adding water vapor to the atmosphere through evaporation from the soils and open water bodies and from the transpiration of water by plants through the stomata in their leaves.

Evergreen. Plants that maintain leaves all year; never without live foliage.

Exotic (species). Nonnative (see also alien); usually refers to introduced animals.

Fauna. All the animal species in a given geographic area.

Flora. All the plant species in a given geographic area.

Forb. Broadleaved, green-stemmed plant; type of herb.

Genus. Taxonomic unit made up of one or more closely related species. The plural is genera.

Geophyte. Perennial plant with its renewal organ protected well below the surface as either a bulb or a corm.

Ground Moraines. Unsorted glacial till materials deposited when glacial ice melts off suddenly and the previously carried materials have to be dropped.

Growthform. Morphology or appearance of a plant that is adapted to particular environmental conditions.

Gullies. Gullies are deep erosional features that started out as small rills.

Habitat. Physical location where a species lives and the local environmental conditions of that place.

Heath. Small-leaved shrubs that are member of the Ericaceae, such as heather, bilberries, and so on; leaves are often drought adapted.

Herb. Nonwoody or soft, green-stemmed plant, may be annual or perennial; broad-leaved herbs are called forbs; grasses are called graminoids.

Humus. Partially decayed biological matter; occurs typically as a brown substance on top of or within the upper soil horizons.

Illuviation. Process of accumulating clay minerals in the soil.

Kettle Holes. Formation is caused when isolated blocks of glacial ice are surrounded by rock debris. When the ice melts, a steep-sided depression is left that will subsequently form a small pond or lake.

Krummholz. Trees typically found near treeline, where they are deformed due to severe weather conditions, particularly extreme wind.

Layering. Sprouting of roots from branches when branches touch the ground.

Leaching. Process of downward movement of dissolved compounds through soil horizons by water.

Lichen. Lifeform composed of a fungus and an alga joined in a symbiotic relationship and classified as a single organism.

Lignified Needles. Coniferous needles whose cell walls are rigid and difficult to be penetrated by decomposition agents.

Moraines. Unsorted glacial debris deposited when glaciers have melted.

Morphology. Structure, form, size, and shape of an organism.

Myccorhiza. Association between certain fungi and the root hairs of many plants.

Myccorhizal Fungi. Fungi associated with plant roots that collect and prepare nutrients for uptake by the plant, and receive carbon in return

Needleleaved. Reference to conifers with slender, pointed leaf shapes; pines, spruces, firs, and so on have such leaves.

Open Canopy. The crown of adjacent plants in the same layer of vegetation that do not touch; sunlight can reach the lower layers or the ground.

Orogenesis. Mountain-building processes.

Orographic. Airmasses are forced to rise and cool as they pass over a mountain barrier; may form clouds and often produce precipitation.

Outwash (glacial). Glacial debris that has been transported and sorted by meltwater from glaciers.

Outwash Plains. Plains built up from glacial outwash materials.

Paludification. Process of developing acidic, waterlogged conditions and thus expanding bogs due to subsequent growth of mosses.

Petiole. Stalk of a leaf.

Photosynthesis. Process in green plants converting oxygen and carbon dioxide in the presence of sunlight to sugars and starches; process releases oxygen.

Physiognomy. Characteristic appearance of a plant association or community.

Pleistocene Epoch. Geologic time scale; 1.6 to 0.01 million years ago.

Pleistocene Ice Ages. Repeated cooling of temperatures during the Pleistocene caused the development of continental ice sheets (glaciers) in the Northern Hemisphere and on alpine mountains. Intermediate warming periods caused the glaciers to melt.

Pliocene Epoch. Geologic time scale; 5.3 to 1.6 million years ago.

Podsolization. Soil-leaching processes in cool, humid climates.

Proteoid Roots. Seasonal roots by plants of the protea family, made up of hundreds of tiny rootlets resembling cotton to facilitate quick gathering of nutrients and moisture after the drought season has ended.

Relict Species. Species that developed in a previous geologic time and have maintained to the present.

Rhizobial Bacteria. Bacteria associated with the roots of some plants; often enclosed in nodules; fix free nitrogen into compounds that plants can use.

Rhizome. Horizontal root structure just below the surface.

Rills. Rills are numerous closely spaced channels when excessive surface erosion occurs. See also gullies.

Salinization. Soluble salts are deposited within the soils when water that has carried these salts evaporates.

Sclerophyll. Tough, drought-resistant thick leaves with a coating or cuticle to protect them from dehydration.

Scrub. Vegetation type characterized by sparse, small shrubs.

Soil. Uppermost terrestrial surface layer; composed of a mixture of mineral and organic matter.

Soil Horizon. Distinctive layer of the soil; differentiated from other layers by color, organic content, mineral composition, particle size, and so on.

Species. Fundamental unit of classification in taxonomy.

Stomata. Small openings on the underside of leaves through which plants exchange gases with the atmosphere.

Succession (ecological). Development of a plant community over time on a barren site. A series of successions develops until the persistent climax is developed that represents an equilibrium with the local environmental conditions.

Successional Stages. Stages in the succession of vegetation communities.

Taxonomy. Science of describing, classifying, and naming organisms.

Tertiary Period. Geologic time scale; 66.4 to 1.6 million years ago.

Till (glacial). Unsorted materials that were directly deposited by a melting glacier.

Tussock. Growthform of sedges and grasses; individual plants grow as tufts or clumps, forming conspicuous hummocks on the ground.

Vascular Plant. Plant with conducting vessels (phloem and xylem) that move nutrients between roots and leaves; includes ferns and flowering plants.

Zonation. Occurrence of a particular lifeform in distinct belts determined by latitude or elevation.

Bibliography

General Works

Breckle, Sigmar-Walter. 2002. *Walter's Vegetation of the Earth*. New York: Springer.

Chapin, F. Stuart, Mark W. Oswood, Keith van Cleve, Leslie A. Viereck, and David L. Verbyla, eds. 2006. *Alaska's Changing Boreal Forest*. New York: Oxford University Press.

FAO. 2000. Global Forest Resources Assessment. http://www.fao.org/forestry/fo/fra/index.html.

FAO. 2001. Global Ecological Zoning for the Global Forest Resources Assessment. http://www.fao.org/docrep/006/ad652e/ad652e00.htm.

Mittermeier, Russell A., Patricio Robles Gil, Michael Hoffmann, John Pilgrim, Thomas Brooks, Cristina Goettsch Mittermeier, John Lamoreux, and Gustavo A. B. Da Fonseca. 2004. *Hotspots Revisited*. Mexico City: CEMEX.

U.S. Defense Mapping Agency. 1992. *Digital Chart of the World*. Washington, DC: U.S. Defense Mapping Agency.

Woodward, Susan L. 2003. *Biomes of Earth: Terrestrial, Aquatic, and Human-Dominated*. Westport, CT: Greenwood Press.

World Wildlife Fund. 2008. Terrestial Ecoregions (reports on individual ecoregions of the world, partially complete). http://www.worldwildlife.org/science/ecoregions.cfm.

Boreal Forests

Bailey, R. G. 1994. *Ecoregions of the United States*. Washington, DC: U.S. Forest Service. http://www.fs.fed.us/land/pubs/ecoregions/ecoregions.html.

Bonan, G. B., and H. H. Shugart. 1989. "Environmental Factors and Ecological Processes in Boreal Forests." *Annual Review in Ecology and Systematics* 20: 128.

Chapin, F. Stuart, Mark W. Oswood, Keith van Cleve, Leslie A. Viereck, and David L. Verbyla, eds. 2006. *Alaska's Changing Boreal Forest*. New York: Oxford University Press.

Christopherson, Robert W. 1994. *Geosystems: An Introduction to Physical Geography*. New York: Macmillan College Publishing.

Elliott-Fisk, Deborah L. 1983. "The Stability of the Northern Canadian Tree Limit." *Annals of the Association of American Geographers* 73: 560–576.

Elliott-Fisk, Deborah L. 2000. "The Taiga and Boreal Forest." In *North American Terrestrial Vegetation*, 2nd ed., ed. Michael G. Barbour and William Dwight Billings, 41–73. Cambridge: Cambridge University Press.

Esseen, Per-Anders, Bengt Ehnstrom, Lars Ericson, and Kjell Sjoberg. 1997. "Boreal Forest." In *Boreal Ecosystems and Landscapes: Structures, Processes, and Conservation of Biodiversity,* ed. Lennart Hannson, 16–47. Copenhagen: Munksgaard International Publishers.

Eyre, S. E. 1968. *Vegetation and Soils*. 2nd ed. Chicago: Aldine Publishing.

FAO. 2001. Global Ecological Zoning for the Global Forest Resources Assessment. http://www.fao.org/docrep/006/ad652e/ad652e00.htm.

Gates, David M. 1993. *Climate Change and Its Biological Consequences*. Sunderland, MA: Sinauer Associates.

Hafen, L. R., W. E. Hollon, and C. C. Rister. 1970. *Western America: The Exploration, Settlement, and Development of the Region beyond the Mississippi*. Englewood Cliffs, NJ: Prentice Hall.

Hare, F. K. 1950. "Climate and Zonal Divisions of the Boreal Forest Formation in Eastern Canada." *Geographical Review* 40: 615–635.

Henry, David J. 2002. *Canada's Boreal Forest*. Smithsonian Natural History Series. Washington, DC: Smithsonian Institution Press.

Houk, Rose. 1993. *Great Smoky Mountains National Park*. New York: Houghton Mifflin.

Johnson, E. A. 1992. *Fire and Vegetation Dynamics: Studies from the North American Boreal Forest*. Cambridge: Cambridge University Press.

Kasischke, Eric S., and Brian J. Stocks. 2000. "Introduction." In *Fire, Climate Change, and Carbon Cycling in the Boreal Forest,* ed. Eric S Kasischke and Brian J. Stocks, 1–6. New York: Springer.

Klinger, Lee F. 1991. "Peatland Formation and Ice Ages: A Possible Gaian Mechanism Related to Community Succession." In *Scientists on Gaia,* ed. S. H. Schneider and P. J. Boston, 247–255. Cambridge, MA: MIT Press.

Künnecke, Bernd H. 1979. *Die kulturlandschaftline Bedeutung des primären Wirtschaftssektors im Staate Oregon, USA*. Regensburg, Germany: Regensburger Geographische Schriften.

Larsen, James A. 1980. *The Boreal Ecosystem*. Physiological Ecology Series. New York: Academic Press.

Larsen, James A. 1982. *Ecology of Northern Lowland Bogs and Coniferous Forests*. New York: Academic Press.

MacDonald, Glen M., Julian M. Szeicz, Jane Claricoates, and Kursti A. Dale. 1998. "Response of the Central Canadian Treeline to Recent Climatic Changes." *Annals of the Association of American Geographers* 88: 183–208.

Nowak, Ronald M. 1991. *Walker's Mammals of the World*. 5th ed., vol. 2. Baltimore: Johns Hopkins University Press.

O'Clair, Rita M., Robert H. Armstrong, and Richard Carstensen. 1992. *The Nature of Southeast Alaska: A Guide to Plants, Animals, and Habitats*. Anchorage: Alaska Northwest Books.

Pastor, J., R. H. Gardner, V. H. Dale, and W. M. Post. 1987. "Successional Changes in Nitrogen Availability as a Potential Factor Contributing to Spruce Decline in Boreal North America." *Canadian Journal of Forest Research* 17: 1394–1400.

Pielou. E. C. 1988. *The World of Northern Evergreens.* Ithaca, NY: Comstock Publishing Associates.

Pielou, E. C. 1991. *After the Ice Age: The Return of Life to Glaciated North America.* Chicago: University of Chicago Press.

Price, Larry W. 1971. "Biogeography Field Guide to the Cascade Mountains: Transect along U.S. Highway 26 in Oregon." Occasional Papers in Geography. Department of Geography, Portland State University, Portland, OR.

Pruitt, W. O., and L. M. Baskin. 2004. *Boreal Forest of Canada and Russia.* Sofia, Bulgaria: Pensoft.

Ricketts, Taylor H., Eric Dinerstein, David M. Olson, Colby J. Loucks, William Eichbaum, Dominick DellaSalla, Kevin Kaunagh, Prashant Hedao, Patrick Hurley, Karen Carney, Robin Abell, and Steven Walters. 1999. *Terrestrial Ecoregions of North America: A Conservation Assessment.* Washington, DC: Island Press.

Rowe, J. S. 1972. "Forest Regions of Canada." Ottawa: Canadian Forestry Service, Department of the Environment.

Runesson, Ulf T. 2007. Boreal Forests: Overview. http://www.borealforest.org/index.php?category=world_boreal_forest&page=overview.

Runesson, Ulf T. 2007. World's Boreal Forests: Animal and Plant Species. http://www.borealforest.org/world/world_species.htm.

Rysin, L. P., and M. V. Nadeshdina, eds. 1968. *Long-Term Biogeocenotic Investigations in the Southern Taiga Subzone.* Jerusalem: Israel Program for Scientific Translations.

Schoonmaker, Peter K., Bettina von Hagen, and Edward C. Wolf. 1997. *The Rain Forests of Home: Profile of a North American Bioregion.* Washington DC: Island Press.

Stanturf, John A., and Palle Madsen, eds. 2005. *Restoration of Boreal and Temperate Forests.* Boca Raton, FL: CRC Press.

Strahler, A. N., and A. H. Strahler. 1989. *Elements of Physical Geography.* 4th ed. New York: John Wiley & Sons.

Thompson, Elizabeth H., and Eric R. Sorenson. 2000. *Wetland, Woodland, Wildland: A Guide to the Natural Communities of Vermont.* Hanover, NH: University Press of New England.

Troll, C., and K. H. Paffen. 1965. "Seasonal Climates of the Earth." In *World Maps of Climatology.* New York: Springer-Verlag.

Vale, Thomas R. 1982. *Plants and People.* Washington, DC: Association of American Geographers.

Van Gelder, Richard G. 1982. *Mammals of the National Parks.* Baltimore: Johns Hopkins University Press.

Vankat, John L. 1979. *The Natural Vegetation of North America.* New York: John Wiley & Sons.

Viereck, Leslie A., and E. L. Little, Jr. 1972. *Alaska Trees and Shrubs.* USDA Forest Service, Agriculture Handbook no. 410. Washington, DC: U.S. Department of Agriculture.

Viereck, Leslie A., and E. L. Little, Jr. 1974. *Guide to Alaska Trees.* USDA Forest Service, Agriculture Handbook no. 472. Washington, DC: U.S. Department of Agriculture.

Walter, Heinrich. 1985. *Vegetation of the Earth and Ecological Systems of the Geo-Biosphere.* 3rd ed. New York: Springer-Verlag.

Woodward, F. I. 1995. "Ecophysiological Controls of Conifer Distributions." In *Ecophysiol-ogy of Coniferous Forests,* ed. William K. Smith and Thomas M. Hinckley, 79–94. San Diego: Academic Press.

Woodward, Susan L. 2003. *Biomes of Earth: Terrestrial, Aquatic, and Human-Dominated.* West-port, CT: Greenwood Press.

Zahn, Ulf, ed. 1996. *Diercke Weltatlas.* 4th ed. Braunschweig, Germany: Westermann.

Temperate Broadleaf Deciduous Forests

Bailey, R. G. 1994. *Ecoregions of the United States.* Washington, DC: U.S. Forest Service. http://www.fs.fed.us/land/pubs/ecoregions/ecoregions.html.

Barnes, Burton V. 1991. "Deciduous Forests of North America." In *Temperate Deciduous Forests,* ed. E. Röhrig and B. Ulrich, 219–344. Ecosystems of the World, 7. Amsterdam: Elsevier.

Braun, E. Lucy. 1950. *Deciduous Forests of Eastern North America.* Philadelphia: Blakiston.

Ching, Kim K. 1991. "Temperate Deciduous Forests in East Asia." In *Temperate Deciduous Forests,* ed. E. Röhrig and B. Ulrich, 539–555. Ecosystems of the World, 7. Amsterdam: Elsevier.

Donoso, Claudio. 1996. "Ecology of *Nothofagus* Forests in Central Chile." In *The Ecology and Biogeography of* Nothofagus *Forests,* ed. Thomas T. Veblen, Robert S. Hill, and Jen-nifer Reed, 271–292. New Haven, CT: Yale University Press.

Eyre, S. R. 1968. *Vegetation and Soils.* 2nd ed. Chicago: Aldine Publishing.

FAO. 2001. Global Ecological Zoning for the Global Forest Resources Assessment. http://www.fao.org/docrep/006/ad652e/ad652e00.htm.

Furley, Peter A., and Walter W. Newey. 1983. *Geography of the Biosphere: An Introduction to the Nature, Distribution, and Evolution of the World's Life Zones.* London: Butterworths.

Goodheart, Adam. 2006. "Smoky Mountain Seasons." *National Geographic* 210 (2): 90–107.

Hanson, Paul J., and Stand Wullschleger, eds. 2003. *North American Temperate Deciduous Forest Responses to Changing Precipitation Regimes.* New York: Springer.

Harshberger, John W. 1970. *The Vegetation of the New Jersey Pine-Barrens.* New York: Dover Publications.

Hill, Robert S., and Mary E. Dettmann. 1996. "Origins and Diversification of the Genus *Nothofagus.*" In *The Ecology and Biogeography of* Nothofagus *Forests,* ed. Thomas T. Veblen, Robert S. Hill, and Jennifer Reed, 11–24. New Haven, CT: Yale University Press.

Houk, Rose. 1993. *Great Smoky Mountains National Park.* New York: Houghton Mifflin.

Jahn, Gisela. 1991. "Temperate Deciduous Forests of Europe." In *Temperate Deciduous Forests,* ed. E. Röhrig and B. Ulrich, 377–502. Ecosystems of the World, 7. Amsterdam: Elsevier.

Kitchings, J. Thomas, and Barbara T. Walton. 1991. "Fauna of the North American Tem-perate Deciduous Forest." In *Temperate Deciduous Forests,* E. Röhrig and B. Ulrich, 345–370. Ecosystems of the World, 7. Amsterdam: Elsevier.

Kricher, John C. 1998. *A Field Guide to Eastern Forests: North America.* Peterson Field Guides. Boston: Houghton Mifflin.

Linzey, Donald W. 1998. *The Mammals of Virginia.* Blacksburg, VA: McDonald & Wood-ward Publishing.

Loucks, Orie. 1998. "In Changing Forests, A Search for Answers." In *An Appalachian Trag-edy: Air Pollution and Tree Death in the Eastern Forests of North America,* ed. Harvard Ayers, Jenny Hager, and Charles Little, 85–97. San Francisco: Sierra Club.

Markgraf, Vera, Edgardo Romero, and Carolina Villagian. 1996. "History and Paleogeography of South American *Nothofagus* Forests." In *The Ecology and Biogeography of* Nothofagus *Forests,* ed. Thomas T. Veblen, Robert S. Hill, and Jennifer Reed, 354–386. New Haven, CT: Yale University Press.

Petersen, Roger T. 1934 (1991 reprint). *A Field Guide to the Birds: Giving Field Marks of All Species Found in Eastern North America.* New York: Houghton Mifflin.

Ricketts, Taylor H., Eric Dinerstein, David M. Olson, Colby J. Loucks, William Eichbaum, Dominick DellaSalla, Kevin Kaunagh, Prashant Hedao, Patrick Hurley, Karen Carney, Robin Abell, and Steven Walters. 1999. *Terrestrial Ecoregions of North America: A Conservation Assessment.* Washington, DC: Island Press.

Röhrig, Ernst. 1991a. "Introduction." In *Temperate Deciduous Forests,* ed. E. Röhrig and B. Ulrich, 1–5. Ecosystems of the World, 7. Amsterdam: Elsevier.

Röhrig, Ernst. 1991b. "Seasonality." In *Temperate Deciduous Forests,* ed. E. Röhrig and B. Ulrich, 25–33. Ecosystems of the World, 7. Amsterdam: Elsevier.

Röhrig, Ernst. 1991c. "Floral Composition and Its Evolutionary Development." In *Temperate Deciduous Forests,* ed. E. Röhrig and B. Ulrich, 17–23. Ecosystems of the World, 7. Amsterdam: Elsevier.

Schaefer, Matthias. 1991. "Fauna of the European Temperate Deciduous Forest." In *Temperate Deciduous Forests,* ed. E. Röhrig and B. Ulrich, 503–525. Ecosystems of the World, 7. Amsterdam: Elsevier.

Schmaltz, Jürgen. 1991. "Deciduous Forests of Southern South America." In *Temperate Deciduous Forests,* ed. E. Röhrig and B. Ulrich, 557–578. Ecosystems of the World, 7. Amsterdam: Elsevier.

Shankman, David, and Justin L. Hart. 2007. "The Fall Line: A Physiographic-Forest Vegetation Boundary." *Geographical Review* 97: 502–519.

Vankat, John L. 1979. *The Natural Vegetation of North America: An Introduction.* New York: John Wiley & Sons.

Veblen, Thomas T. 2007. "Temperate Forests of the Southern Andean Region." In *The Physical Geography of South America,* ed. Thomas T. Veblen, K. R. Young, and A. R. Orme, 217–231. Oxford: Oxford University Press.

Veblen, Thomas T., Claudio Donoso, Thomas Kitberger, and Alan J. Rebertus. 1996. "Ecology of Southern Chilean and Argentinean *Nothofagus* Forests." In *The Ecology and Biogeography of* Nothofagus *Forests,* ed. Thomas T. Veblen, Robert S. Hill, and Jennifer Reed, 293–353. New Haven, CT: Yale University Press.

Veblen, Thomas T., K. R. Young, and A. R. Orme, eds. 2007. *The Physical Geography of South America.* Oxford: Oxford University Press.

Whitney, Gordon G. 1994. *From Coastal Wilderness to Fruited Plain: A History of Environmental Change in Temperate North America, 1500 to the Present.* Cambridge: Cambridge University Press.

Wieboldt, Thomas F. 1989. "Early Botanical Exploration and Plant Notes from the New River, Virginia, Section." In *Proceedings of the Eighth Annual New River Symposium,* 149–159. Oak Hill, WV: National Park Service.

Williams, M. 1989. *Americans and Their Forests: A Historical Geography.* Cambridge: Cambridge University Press.

Ziegler, Susy Svatek, 2007. "Postfire Succession in an Adirondack Forest." *Geographical Review* 97: 467–483.

Mediterranean Woodland and Scrub

Arianoutsou, M., and R. H. Groves, eds. 1994. *Plant-Animal Interactions in Mediterranean-Type Ecosystems.* Dordrecht: Kluwer Academic Publishers.

Axelrod, D. I. 1989. "Age and Origin of Chaparral." In *The California Chaparral: Paradigms Revisited*, ed. S. C. Keeley, 7–19. Los Angeles: Natural History Museum of Los Angeles County.

Bahre, Conrad J. 1979. *Destruction of the Natural Vegetation of North-Central Chile.* Berkeley: University of California Press.

Blondel, Jacques, and James Aronson. 1999. *Biology and Wildlife of the Mediterranean Region.* Oxford: Oxford University Press.

Bond, P., and P. Goldblatt. 1984. "Plants of the Cape Flora: A Descriptive Catalogue." *Journal of South African Botany,* supp. vol.

Conacher, Arthur J., and Maria Sala, eds. 1998. *Land Degradation in Mediterranean Environments of the World: Nature and Extent, Causes and Solutions.* New York: John Wiley & Sons.

Cowling, R. M., D. M. Richardson, and S. M. Pierce. 1997. *Vegetation of South Africa.* Cambridge: Cambridge University Press.

Dallman, Peter. R. 1998. *Plant Life in the World's Mediterranean Climates.* Oxford: Oxford University Press.

Davis, G. W., and D. M. Richardson, eds. 1995. *Mediterranean-Type Ecosystems: The Function of Biodiversity.* New York: Springer-Verlag.

di Castri, F., and Harold A. Mooney, eds. 1973. *Mediterranean-Type Ecosystems: Origin and Structure.* New York: Springer-Verlag.

FAO. 2001. Global Ecological Zoning for the Global Forest Resources Assessment. http://www.fao.org/docrep/006/ad652e/ad652e00.htm.

Furley, Peter A., and Walter W. Newey. 1983. *Geography of the Biosphere: An Introduction to the Nature, Distribution, and Evolution of the World's Life Zones.* London: Butterworths.

Groves, R. H., and F. di Castri, eds. 1991. *Biogeography of Mediterranean Invasions.* Cambridge: Cambridge University Press.

Kalin-Arroyo, M. T., P. H. Zedler, and M. D. Fox, eds. 1995. *Ecology and Biogeography of Mediterranean Ecosystems in Chile, California, and Australia.* New York: Springer-Verlag.

Kingdon, Jonathan. 1989. *Island Africa: The Evolution of Africa's Rare Animals and Plants.* Princeton, NJ: Princeton University Press.

Low, A. Barrie, and A. G. Rebelo, eds. 1996. *Vegetation of South Africa, Lesotho, and Swaziland.* Pretoria: Department of Environmental Affairs and Tourism. http://www.ngo.grida.no/soesa/nsoer/Data/vegrsa/vegstart.htm.

Mittermeier, Russell A., Norman Myers, and Cristina Goettsch Mittermeier. 1999. *Hotspots: Earth's Biologically Richest and Most Endangered Terrestrial Ecosystems.* Mexico City: CEMEX.

Mittermeier, Russell A., Patricio Robles Gil, Michael Hoffmann, John Pilgrim, Thomas Brooks, Cristina Goettsch Mittermeier, John Lamoreux, and Gustavo A. B. Da Fonseca. 2004. *Hotspots Revisited.* Mexico City: CEMEX.

Mittermeier, Russell A., Cyril F. Karmos, Cristina Goettsch Mittermeier, Patricio Robles Gil, Trevor Sandwith, and Charles Besancon. 2005. *Transboundary Conservation: A New Vision for Protected Areas.* Mexico City: CEMEX.

Moreno, J. M., and W. C. Oechel, eds. 1994. *The Role of Fire in Mediterranean-Type Ecosystems*. New York: Springer-Verlag.

Moreno, J. M., and W. C. Oechel, eds. 1995. *Global Change and Mediterranean-Type Ecosystems*. New York: Springer-Verlag.

Ricketts, Taylor H., Eric Dinerstein, David M. Olson, Colby J. Loucks, William Eichbaum, Dominick DellaSalla, Kevin Kaunagh, Prashant Hedao, Patrick Hurley, Karen Carney, Robin Abell, and Steven Walters. 1999. *Terrestrial Ecoregions of North America: A Conservation Assessment*. Washington, DC: Island Press.

Rundel, P. W., G. Montenegro, and F. M. Jaksic, eds. 1998. *Landscape Disturbance and Biodiversity in Mediterranean-Type Ecosystems*. Berlin: Springer.

Thrower, N. J. W., and D. E. Bradbury. 1977. *Chile-California Mediterranean Scrub Atlas: A Comparative Analysis*. Stroudsburg, PA: Dowden, Hutchinson, & Ross.

Veblen, Thomas T., K. R. Young, and A. R. Orme, eds. 2007. *The Physical Geography of South America*. Oxford: Oxford University Press.

White, Mary E. 1990. *The Flowering of Gondwana: The 400 Million Year Story of Australian Plants*. Princeton, NJ: Princeton University Press.

Index

About the Author

BERND H. KUENNECKE is professor and chair of the Department of Geography at Radford University in Virginia, as well as director of the university's GIS Center. His undergraduate work was done at the Universität Regensburg in Germany, followed by graduate work at the University of Oregon with a thesis on agricultural irrigation in Oregon. His Ph.D. in geography was completed at the Universität Regensburg. He has been a member of the teaching faculty at Radford University since 1977. His teaching duties over the past thirty-one years have included courses in physical geography, economic geography, planning, field research methods, GIS, land use planning, digital image processing, perspectives in geography, North America and world regional courses, and directed field research courses. He has taken numerous research and teaching trips with and without university-level students through the temperate forest regions of the United States and Canada (including several trips to Alaska via various routes), and he has traveled extensively throughout the temperate forests of Europe. He has worked and published on the conversion of the natural environments of Oregon through primary economic activities.